Asterisk 1.4

The Professional's Guide

Implementing, administering, and consulting on
commercial IP telephony solutions

Colman Carpenter

David Duffett

Nik Middleton

Ian Plain

BIRMINGHAM - MUMBAI

Asterisk 1.4
The Professional's Guide

First published: August 2009

Production Reference: 1030809

Published by Packt Publishing Ltd.
32 Lincoln Road
Olton
Birmingham, B27 6PA, UK.

ISBN 978-1-847194-38-1

www.packtpub.com

Cover Image by Vinayak Chittar (vinayak.chittar@gmail.com)

Credits

Authors
Colman Carpenter
David Duffett
Nik Middleton
Ian Plain

Reviewers
Ian Plain
Jared Smith
Philippe Lindheimer

Acquisition Editor
James Lumsden

Technical Editors
Gagandeep Singh
Charumathi Sankaran

Indexer
Rekha Nair

Editorial Team Leader
Gagandeep Singh

Project Team Leader
Lata Basantani

Project Coordinator
Neelkanth Mehta

Proofreader
Laura Booth

Production Coordinator
Aparna Bhagat

Cover Work
Aparna Bhagat

Foreword

Watching Asterisk move from being a personal coding project to a community of tens of thousands of programmers and millions of users has been quite the ride so far! Asterisk is only now hitting its prime, and there are so many more things that creative people are going to do with the code. The growth of the project over the years has stunned and pleased me, and it's amazing that well-written and comprehensive books like this now exist to help more advanced users navigate the waters of larger and more complex Asterisk installations. Asterisk installations are now huge, both in numbers of locations and the unimaginably large size of many of those locations — thousands or tens of thousands of users! Asterisk implementations are rarely limited by the capability of the software but more often by not knowing how to utilize it. Books like this play an important role in getting the experience of those who have already done in the hands of those who want to do.

Hopefully the knowledge here allows you to continue your adventure with Asterisk, moving from the basics of PBX construction to having the ability to quickly implement advanced call logic processes and work with the more exotic telephony and VoIP interfaces. The motto of "There's more than one way to do it!" is almost always true with Asterisk — this book seems to contain an excellent cross-section of at least one of those ways to do "it" (whatever "it" happens to be for your application) and you'll quickly think of many other ways once you've mastered the methods shown.

The authors here have really shown some excellent detailed explanations of how to use Asterisk, and I hope this provides the incentive for you, the reader, to experiment in more wide-ranging ways with Asterisk once you've understood the basics. Most of the Asterisk community has learned with hands-on experimentation, and it's great to see more encouragement of this type of learning as is contained in these pages. Kudos to the authors, especially David Duffett, who has been involved with Asterisk for so long and has taught so many people their first dialplan routines (and hopefully has left them uninjured from his famous habit of throwing candy at people who give correct answers in class or in his talks).

Soon you'll be doing least-cost-routing, integrating your instant messenger system with your mobile phone calls, controlling robots with voice commands via your phone, or dreaming up a new company based on some voice-based service that nobody has tapped into yet. And the best thing about Asterisk is that it remains open source—if you come up with a feature or enhancement that you think *must* be in Asterisk, then the good news is that it *can* be! Become a member of the Asterisk community, and your contributed code could be included. We all anxiously await your book, your product, or just your involvement with the Asterisk community.

Mark Spencer
Chairman and CTO of Digium

About the Authors

Colman Carpenter is the MD of Voicespan, a Kent-based company that offers Asterisk-based systems to the SME market across the UK. He is an IT professional of over 20 years standing, with experience in diverse areas such as IBM mid-range software development, Lotus Notes and Domino consultancy, Data Management, E-marketing consultancy, IT Management, Project Management, Wordpress Website Design, and lately, Asterisk consultancy. He is a qualified PRINCE2 practitioner.

Voicespan (`http://www.voicespan.co.uk`) offers Asterisk-based systems as the cornerstone of a holistic VoIP-telephony service for SMEs. They offer companies a one-stop shop for implementing a VoIP-capable system, encompassing Asterisk-based systems, endpoints, trunks, telephony interfaces and network equipment, and the consultancy necessary to bring it all together into a coherent whole. This is his first book.

> I would like to thank my wife, Hazel, and daughters, Caiti and Fay, for their support during the writing of this book. At times it seemed like you believed more than I in my ability to do so!

David Duffett delivers Asterisk training and consultancy around the world through his own company (TeleSpeak Limited, `www.telespeak.co.uk`), in addition to designing and delivering training for a number of companies, including Digium, Inc.

A keen Asterisk enthusiast, David also enjoys podcasting, radio presenting, and teaching public-speaking skills. He is a Chartered Engineer with experience in fields including Air Traffic Control communications, Wireless Local Loop, Mobile Networks, VoIP, and Asterisk. David has been in the telecoms sector for nearly 20 years and has had a number of computer telephony, VoIP, and Asterisk articles published through various industry publications and web sites.

Nik Middleton has been in wide-area communications since the mid-eighties. He spent most of the nineties working in the US, where he developed a shareware Microsoft mail to SMTP/POP3 connector that sold some 287,000 copies. He spent six years working for DuPont in VA, developing remote monitoring systems for their global Lycra business. In late 2000, he returned to the UK where he held various senior positions in British Telecom, LogicaCMG, and Computer Science Corp.

In 2005, tired of working in London, he set up his own company (Noble Solutions) providing VoIP solutions in rural Devon, where he now lives with his wife Georgina and three children, Mathew, Vicky, and Isabel. A keen amateur pilot, his favorite place when not in the office is flying over the beautiful Devon countryside.

Ian Plain has worked in the telecoms industry since 1981 and has designed some of the largest PBX networks in the UK. Since the late 1990s, he has been involved with VoIP initially for links between systems, and with IP PBX systems since 1999. Since 2003, he has been running a telecoms consultancy based near Bath in the UK, working primarily on high-availability Asterisk-based solutions for corporate customers.

About the Reviewers

Ian Plain: Please see the entry in *About the Authors*.

Jared Smith is the Training Manager for Digium, Inc. As a long time Asterisk user, contributor, and evangelist, he has spent the last several years helping the Asterisk community. Jared is a dynamic and knowledgeable instructor with several years of experience in leading various Asterisk training classes.

He is also co-author of *Asterisk: The Future of Telephony*, *O'Reilly Media* and regularly writes other Asterisk documentation as well.

Jared holds a Bachelors of Science degree in Computer Engineering from the Utah State University and currently lives in Virginia with his wife and two children.

Philippe Lindheimer is the project leader and primary developer of FreePBX and serves as the Open Source Community Director at Bandwidth.com, the corporate sponsor of the FreePBX project (the most widely deployed Asterisk-based PBX/ GUI open-source application in the world). He cofounded and runs the Open Telephony Training Seminar providing FreePBX/Asterisk technical and marketing training to resellers and end users. Originally with Hewlett Packard, he has been in the engineering industry for over two decades, working on a range of technical consulting roles with many Fortune 500 Companies.

He has a BS (Hons) in EE/CS from the University of Colorado, Boulder. He now lives in the Seattle, WA area.

Table of Contents

Preface

This book is a sequel to *Building Telephony Systems with Asterisk*, which started you on a journey to the summit of Asterisk knowledge, taking you from base camp to camp two, from being a complete Asterisk newbie to a competent telephony system builder and manager. Now it's time to push to the top, to take your telephony knowledge to a point where you can build high-performance, resilient, and professional PBXs using the most popular open source telephony software in the world — Asterisk.

In that book, the focus was very much on installing and configuring Asterisk for a number of common scenarios, including both home and office use. This it achieved admirably, so you may now wonder why another book is needed. Well, there are three main reasons for writing this book. Firstly, Asterisk is such a highly-capable and configurable telephony engine that the 150-odd pages in the book necessarily had to exclude discussion of some of the more advanced features, which we now have the opportunity to explore. Secondly, Asterisk is invariably implemented as part of an IP network, and further examination of network considerations is warranted. Finally, like all popular open source software, Asterisk is constantly being updated, and while this book still assumes the version 1.4 of Asterisk is in use, we do point out any differences in version 1.6 where relevant, such as the change from Zaptel to DAHDI.

Therefore, the goal of this book is to give you enough knowledge to build and install a telephony system with Asterisk at its core, which will stand comparison with the market-leading commercial IP-enabled systems. Whether you are building such a system as a result of an internal company requirement, or you plan to offer it as an element of a commercial package to customers, this book will take you through all the areas that require consideration. On reading this book you will also be in a position to understand the real-life issues you are likely to experience when deploying such a system, both technical and otherwise.

By its very nature, Asterisk demands that much of the focus of this book be on the technical aspects of building your professional system. However, as with most IT implementations, success will also rely on "soft" issues such as managing expectations, understanding and meeting the customer's particular needs, and ensuring delivery is on time and up to the budget. Hence, where appropriate, we make mention of the non-technical aspects that may make a difference to your deployment.

To achieve our goal, this book will build on knowledge already gained by reinforcing that learning and adding extra skills covering:

- Security
- Networks
- Large-scale considerations
- Resilience
- Scalability
- Integration with complementary products
- Commercial aspects

Reviewing the basics

If you have not already done so, it is recommended that you read *Building Telephony Systems with Asterisk*, or achieve a good degree of competence in building basic Asterisk PBXs through other means. These could include commercial training courses (see www.digium.com/en/training for further details) or openly available internet resources such as the excellent VoIP wiki at http://www.voip-info.org.

While most people with a day-to-day exposure to Asterisk systems should stand to gain much from this book, it has been written in the expectation that you will possess the following Asterisk skills and experience, ideally gained through text file configuration:

- Connecting Asterisk to analogue and digital PSTN lines, and VoIP services
- Configuring different types of terminal equipment (phones, communication devices, other PBXs)
- Installing Asterisk, Zaptel and LibPRI
- Configuring features (Voicemail, Music On Hold, Queues, Conference Rooms, and so on)

- Creating a dialplan, including call distribution
- CDRs, call monitoring and recording
- Backups and restores
- Basic security and load balancing

Once equipped with this knowledge you stand to gain the maximum from the topics covered in this book, enabling you to build professional Asterisk systems to be deployed internally, or to form the cornerstone of a commercial offering.

No compromise

In this book you will, hopefully, learn many new things. At its conclusion you will have the knowledge to build and successfully implement systems that combine great performance, resilience and stability. In order to do so, we will mainly consider "pure" Asterisk systems that require a deep understanding of the dialplan and configuration files without the safety-net of a GUI in between. Think of it as learning to become a great car mechanic. You can certainly be a good mechanic earning a good living by learning how to use a laptop plugged into an engine management system. But if you want to take that extra step to being a true master of the trade then you need to understand at a very deep level just how the internal combustion engine works. So it is with Asterisk. It is perfectly feasible to put very good solutions together using GUI-based systems such as the Digium-owned Switchvox, Trixbox (formerly `Asterisk@Home`) or PBX in a Flash, but to construct the best systems you will need to understand what is happening "under the hood" so that you can tweak them appropriately to achieve or exceed the customers' expectations.

One advantage of eschewing the GUI approach is a potential increase in performance and scalability through the use of a highly-optimized dialplan and a reduction in applications running on the server. However, there are many situations where a GUI is at least as appropriate, particularly if the customer wishes to carry out day-to-day management tasks. Therefore, in Chapter 12 we look at the implications of choosing a GUI-based solution over a "vanilla" system.

[To follow the "trusted network" of Asterisk developers please visit: `www.asteriskpro.co.uk`]

What this book covers

As a result of reading this book, you can expect to build on existing knowledge and gain new skills. Each chapter covers a particular topic, but throughout there is a focus on building an Asterisk system that can form the cornerstone of a serious commercial product, capable of matching or even exceeding the performance of well-known licensed products.

Chapter 1 talks about dialplan techniques including modular implementations by using macros, contexts, and so on to both refine the dialplan and improve the security of the system. It also discusses the use of the `devstate()` function.

Chapter 2 discusses customer network requirements and offers some good advice about potential issues within the customer network and how to resolve them, including the use of VLANs and Quality of Service.

Chapter 3 looks at routing in general, including Least Cost Routing (local, national, and international GSM gateways), fall-back routing, alternate routing, and so on. ENUM and DUNDi are also explained within this context.

Chapter 4 considers call center requirements, including queues, agents, call distribution strategies, performance monitoring and call recording issues. An Asterisk-based call center solution, VICIDIAL, is also discussed in some detail.

Chapter 5 introduces speech technology in the form of ASR, TTS, and SVI; followed by implementation advice and examples. Both Lumenvox and Cepstral packages are explored in detail.

Chapter 6 looks at methods that can be used to implement call accounting and billing solutions for Asterisk. In particular, Asterisk-stat and A2Billing are explored.

Chapter 7 discusses resilience and stability, giving you a guide to implementing highly-available Asterisk solutions for mission-critical applications. Use of failover and load-balancing techniques are explored.

Chapter 8 explores the comprehensive localization options within Asterisk, and also suggests some easily deployed security measures.

Chapter 9 considers interfaces with traditional analogue and digital telephony, giving more in-depth explanations of Libpri and DAHDI (formerly Zaptel), and discussing implementation considerations.

Chapter 10 tackles the good and bad points of using wireless technologies with Asterisk, covering Wi-Fi, dual-mode and DECT handsets. Some suggestions on routing via cell/mobile networks are also offered.

Chapter 11 looks at the good and bad points of Asterisk Graphical User Interfaces (GUIs), focusing on one of the most popular incarnations, FreePBX.

In *Appendix A* we also explore some of the **softer** skills required when selling Asterisk-based solutions, suggesting some sales strategies that can help you in a commercial environment.

In *Appendix B* you will find information you might want to include in sample emails when pitching.

In *Appendix C* you will find a sample appointment sheet which can be used as a template.

Onwards

So now our campsite has been packed away and it is time for the next part of our journey to begin, for those first purposeful steps to be taken towards the summit. We will start in Chapter 1 by looking at the heart of any Asterisk system, the dialplan. You will already have significant knowledge in this area, but we are about to show you some of the techniques that are used in systems with thousands of extensions that handle many tens of thousands of calls per day. Without these techniques, a dialplan can become an unholy mess as system size increases. However, using these techniques will ensure that complexity is avoided and performance is maintained.

Conventions

In this book, you will find a number of styles of text that distinguish between different kinds of information. Here are some examples of these styles, and an explanation of their meaning.

Code words in text are shown as follows: "We can include other contexts through the use of the `include` directive."

A block of code is set as follows:

```
exten => s,1,Dial(Zap/1,30)
exten => s,n,Goto(s-${DIALSTATUS},1)
exten => s,n,Hangup()
exten => s-NOANSWER,1,Voicemail(100,u)
exten => s-BUSY,1,Voicemail(100,b)
exten => i,1,Voicemail(0,s)
```

When we wish to draw your attention to a particular part of a code block, the relevant lines or items are set in bold:

```
[default]
exten => s,1,Dial(Zap/1|30)
exten => s,2,Voicemail(u100)
exten => s,102,Voicemail(b100)
exten => i,1,Voicemail(s0)
```

Any command-line input or output is written as follows:

```
# cp /usr/src/asterisk-addons/configs/cdr_mysql.conf.sample
   /etc/asterisk/cdr_mysql.conf
```

New terms and **important words** are shown in bold. Words that you see on the screen, in menus or dialog boxes for example, appear in the text like this: "clicking the **Next** button moves you to the next screen".

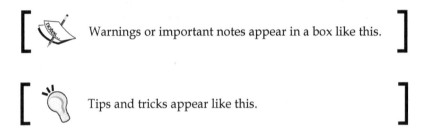

Warnings or important notes appear in a box like this.

Tips and tricks appear like this.

Reader feedback

Feedback from our readers is always welcome. Let us know what you think about this book—what you liked or may have disliked. Reader feedback is important for us to develop titles that you really get the most out of.

To send us general feedback, simply send an email to feedback@packtpub.com, and mention the book title via the subject of your message.

If there is a book that you need and would like to see us publish, please send us a note in the **SUGGEST A TITLE** form on www.packtpub.com or email suggest@packtpub.com.

If there is a topic that you have expertise in and you are interested in either writing or contributing to a book on, see our author guide on www.packtpub.com/authors.

Customer support

Now that you are the proud owner of a Packt book, we have a number of things to help you to get the most from your purchase.

Downloading the example code for the book

Visit http://www.packtpub.com/files/code/4381_Code.zip to directly download the example code.

The downloadable files contain instructions on how to use them.

Errata

Although we have taken every care to ensure the accuracy of our content, mistakes do happen. If you find a mistake in one of our books—maybe a mistake in the text or the code—we would be grateful if you would report this to us. By doing so, you can save other readers from frustration, and help us to improve subsequent versions of this book. If you find any errata, please report them by visiting http://www.packtpub.com/support, selecting your book, clicking on the **let us know** link, and entering the details of your errata. Once your errata are verified, your submission will be accepted and the errata added to any list of existing errata. Any existing errata can be viewed by selecting your title from http://www.packtpub.com/support.

Piracy

Piracy of copyright material on the Internet is an ongoing problem across all media. At Packt, we take the protection of our copyright and licenses very seriously. If you come across any illegal copies of our works, in any form, on the Internet, please provide us with the location address or website name immediately so that we can pursue a remedy.

Please contact us at copyright@packtpub.com with a link to the suspected pirated material.

We appreciate your help in protecting our authors, and our ability to bring you valuable content.

Questions

You can contact us at questions@packtpub.com if you are having a problem with any aspect of the book, and we will do our best to address it.

1
The Dialplan

The **dialplan** is the routing core of an Asterisk server. Its sole role is to look at what is dialed, and route the call to its destination. This is the core of any telephony system and Asterisk is no different.

The dialplan is made up of three elements—extensions, contexts, and priorities. An **extension** is number or pattern that the dialed number is to be matched against and a **context** is a collection of extensions (and possibly other included contexts too). Each extension will have one or more priorities, each of which appear on a separate line, and the priority sequence always starts with the priority "1".

If you have read *Building Telephony Systems with Asterisk*, you will know how to use extensions, priorities, contexts, and included contexts to handle incoming and outgoing calls as well as to set up features such as:

- Call Queues
- Call Parking
- Direct Inward Dialling
- Voicemail
- Automated Phone Directory
- Conference Rooms

In this chapter, we will build on this knowledge by looking at:

- Significant updates since Asterisk 1.2
- Pattern ordering within and between contexts
- Extending the dialplan with variables
- The DEVSTATE() function
- The SYSTEM application

We will then use this knowledge to provide examples of:

- Advanced call routing with the DEVSTATE() function
- Call routing based on the time of the day
- Using multiple ADSL lines within Asterisk to boost call capacity

Dialplan location

The dialplan is primarily defined in the extensions.conf file. This can also include additional files that are added into it using the #include directive. For instance, systems using the FreePBX GUI will have extensions_additional.conf, extensions_custom.conf, and extensions_override_freepbx.conf as standard files, which have been added using #include into the extensions.conf file. We must also remain aware of files such as the features.conf file, as they also include numbers that can be dialed such as codes for Pickup and Call Parking, and so form part of the dialplan.

A list of standard and optional Asterisk configuration files can be found at http://www.voip-info.org/wiki/view/Asterisk+config+files.

Extensions and contexts

Being familiar with Asterisk, you will have a good working understanding of extensions and contexts already. They are, of course, the very heartbeat of Asterisk, and as such they are probably subject to the most change from version to version, as Asterisk evolves to cater for new hardware, software, and more complex working practices. So let's have a quick review of extensions and contexts, pointing out significant changes in versions 1.4 and 1.6, before we proceed to the more advanced techniques and uses.

Pattern matching

Within the dialplan, matching can be either direct or partial against a pattern. Normally in a PBX, these patterns are numeric. But with Asterisk, they can also be alphanumeric or even just alpha. For example 2000, DID01234123456, and Main_number are all valid extensions. As very few phones contain alphabetic keys, the last two are typically only used for incoming DID channels. For the majority of this chapter, we will stick to numeric patterns.

Let's start to explore pattern matching by looking at an extremely simple dialplan:

```
[context_1]
exten => 123,1,Answer()
exten => 123,n,SayDigits(999${CALLERID(num)})
exten => 123,n,Hangup()
```

In this dialplan, when a user with a context of `context_1` dials `123`, they will hear 999 and their caller ID will be read back to them.

Now let's look at a slightly more complex context:

```
[context_1]
exten => _1X.,1,Answer()
exten => _1X.,n,SayDigits(${EXTEN}${CALLERID(num)})
exten => _1X.,n,Hangup()
exten => 123,1,Answer()
exten => 123,n,SayDigits(123${CALLERID(num)})
exten => 123,n,Hangup()
```

You might expect that `123` would match against the `_1X.` extension, as that appears first in the context. However, the way Asterisk orders the dialplan when loading means that exact matches are checked for before pattern matches. Hence if you dial 123, it matches against the 123 pattern first and not the `_1X.` pattern. This pattern would only route the call if an exact match did not exist in the context.

 It is sensible not to use the pattern `_.` as a catch-all pattern, as this will catch the Asterisk special extensions like i, t, h as well. It is far better to use the `_X` pattern.

Once understood, pattern matching is pretty straightforward and does what we expect. However, if you introduce included contexts into the mix, things may work in a way you did not expect and the order needs to be thought through carefully. In particular, it's crucial to understand that Asterisk only checks included contexts after checking for exact matches *and* pattern matches in the local context. The following example illustrates this:

```
[context_1]
include => context_2
exten => _1X.,1,Answer()
exten => _1X.,n,SayDigits(${EXTEN}${CALLERID(num)})
exten => _1X.,n,Hangup()
include => context_3
exten => 123,1,Answer()
exten => 123,n,SayDigits(123${CALLERID(num)})
exten => 123,n,Hangup()
```

The above dialplan is sorted internally by Asterisk shown as follows, and you can see that though the included contexts are at the top and in the middle, the local context is read first, *then* the included contexts are read in the order that they were added. Hence, in this case, a dial string of 122 would be matched by the _1X. pattern before the included contexts are searched.

```
'123'  =>             1. Answer()
                      2. SayDigits(123${CALLERID(num)})
                      3. Hangup()
'_1X.' =>             1. Answer()
                      2. SayDigits(${EXTEN}${CALLERID(num)})
                      3. Hangup()
Include =>            'context_2'
Include =>            'context_3'
```

> If you have a catch-all pattern in your dialplan, consider putting it into a separate context. You can then use the include directive to append that context to the end of the active context, thus ensuring that all of the other pattern matching is attempted first.

One of the most powerful tools you will use on the Asterisk command line is dialplan show <exten>@<context>. For example:

dialplan show 122@context_1

This will show you the matching order that Asterisk will use for the given extension in the specified context, and if there are matches in any included contexts, those contexts will be explicitly identified.

Finally, in a context you may have a switch statement, which includes the dialplan of an external system into the local dialplan. In essence, it's an include for remote systems. Though typing dialplan show will always show the switch statement at the bottom, the defined context on the remote system is searched *after* the local context on your system and *before* any local included contexts! So again, you have to be very careful as to what is the context on the remote system as this will be searched before your included contexts.

The syntax of the switch state is as follows:

```
switch =>IAX2/user:[key]@server/context
```

The user and key are defined in the called server's iax.conf file, and the context is, of course, in the server's dialplan.

Why use contexts?

In our examples so far we could have achieved the desired results very easily without the use of multiple contexts. The simple functionality we have looked at could be carried out in a single, all-encompassing context. In practice, this approach could be applicable for systems with a very limited number of users and trunks, and with very restricted functionality, as there may not be a need to restrict the calling habits of a subset of users.

Use of contexts becomes desirable when we need to offer different options to different users. This is likely to be most applicable in medium and large companies, where you may have "users" ranging from the CEO down to an emergency phone in a lift. However, it can also be the case in smaller companies, where you might want to restrict home workers from making international calls for instance. When you get many different types of users, writing a distinct dialplan for each becomes problematic. The sheer size and complexity of the dialplan will make code management very complicated.

To simplify things, we first need to think about what makes the dialplan for each extension different. Then we need to think about *what remains the same for each extension*, as this needs to be made to work as well. What we often find is that most of these differences can be stored and called in two main ways:

- The user's context
- Variables linked to that user

We will come to variables shortly, but the grouping of extensions into contexts allows us to separate concise and distinct functions from each other. In doing so, we can control very tightly which contexts are used in each scenario, and also implement one "master" copy of each distinct function, aiding maintenance of the code.

Call barring made simple

To illustrate, let's expand our context a bit and use call barring as an example. We will initially have three levels for this example—**local**, **national**, and **international**.

These are defined as follows:

- Any number starting with a 1-9 is local
- Anything starting with a 00 is international.
- Anything else starting with a 0 is national or a mobile number.

This is a simplified example, and uses the UK format of dial prefixes.

We have in this example three contexts — local_num, national_num and international_num. These would correspond to the levels of access we have decided on for our users. For example, an executive phone would be allowed access to all numbers whereas a phone on the shop floor may only be allowed access to local numbers.

We will create the three contexts shown as follows. All we are doing in our example is reading back 1, 2, or 3 to indicate the pattern that has been matched followed by the number dialed—${EXTEN}.

```
[local_num]
Exten => _Z.,1,Answer()
Exten => _Z.,n,SayDigits(1${EXTEN})
Exten => _Z.,n,Hangup()
;
[national_num]
Exten => _0Z.,1,Answer()
Exten => _0Z.,n,SayDigits(2${EXTEN})
Exten => _0Z.,n,Hangup()
;
[international_num]
Exten => _00X.,1,Answer()
Exten => _00X.,n,SayDigits(3${EXTEN})
Exten => _00X.,n,Hangup()
```

For each context we could write an ordered list to cover all patterns, but it is much neater to create a master context for each user. For example:

```
[local]
Include => local_num

[national]
Include => national_num
Include => local_num

[international]
Include => international_num
Include => national_num
Include => local_num
```

Therefore, in the previous example, a user with the national context can dial a normal national number, but not an international number. A user with the international context has the ability to dial both numbers.

This is a pretty simple example with just three level of access, but the modular nature due to the use of contexts allows us to expand it very quickly and easily. For example, we have a user 1000 (our CEO) and he can dial internationally. We also have 1098 and 1099, which are users on the shop floor, and can dial reception and the emergency services.

In this example, we give our CEO a context of [supauser],while the shop floor has a context of [emergencyuser].

The [supauser] context has to be able to dial everything, so it looks like this:

```
[supauser]
include => premium_num ; allows dialing to premium rate numbers
include => international_num ; allows international dialing
include => national_num ; allows national calls
include => mobile_num ; allows calls to mobile phones
include => local_num ; allows local rate calls
include => free_num ; allows free calls such as 800 or
operator services
include => internal_num ; allows the calling of extensions
include => emergency ; allows calls to the emergency services
include => default ; allows access to system features
```

The shop floor just has the following context:

```
[emergencyuser]
include => emergency ; allows calls to emergency services reception.
```

As you can see, we can mix and match these contexts to cover many different types of extensions. Although you may be asking, "Will this really save me time?" well, let's look at two examples. Firstly, our supplier reduces the cost of UK 0870 numbers to free in the evenings as has happened in the UK with BT(British Telecom). Secondly, we also want the shop floor phone (1099) to be able to dial extensions and toll free calls, but not change the dialplan for 1098.

We will deal with the simplest of these extensions (1099) first. All we need to do is change the context associated with this user to a new context called [freeuser]:

```
[freeuser]
include => free_num ; allows calls to free numbers
include => internal_num ; allows the calling of extensions
include => emergency ; allows calls to the emergency services
include => default ; allows access to system features .
```

This is a fast and easy change, which will have no effect on other shop floor users.

And to the change to 0870 numbers, this once again can be put into effect very simply. The only change is that evening and weekend calls are now free. Therefore, we could put it into a [free] context. Although, it isn't always free. It is free only at weekends which would not be suitable. Hence, for this we use the GotoIfTime application, which sets the context, extension, and priority in the channel based on the system time, day, date, and month supplied by the OS.

By adding the following to the free context, users can now dial 0870 numbers at the defined times.

```
exten => _0870XXXXXXX,1,GotoIfTime(17:59-08:00,mon-fri,*,*?national,
${EXTEN},1)
exten => _0870XXXXXXX,1,GotoIfTime(*,sat-sun,*,*?national,${EXTEN},1)
```

In this case, we have made a change for all users who also have a context allowing both local and free calls (as their context includes the free context).

Time and day call routing

The GotoIfTime() application can introduce some powerful functionality into your dialplan if used properly. An example that follows is for a support company where calls are routed to the call centre or staff member on call at a specific time. The customer had centers round the globe and we routed the calls to whichever center was open at that time of day.

```
[folthesun]
;
;This section sets the constants and variables for numbers and times
;Nine timezones are defined to allow for 4 a day and sat and sun
working
;At present there are 6 destinations for NA AU and EMEA
;
exten => s,1,set(__tzone1=00:00-07:59)
exten => s,n,set(__tzone2=08:00-17:30)
exten => s,n,set(__tzone3=17:31-23:59)
exten => s,n,set(__tzone4=17:31-23:59)
exten => s,n,set(__tzone5=00:00-23:59)
exten => s,n,set(__tzone6=00:00-23:59)
exten => s,n,set(__tzone7=00:00-23:59)
exten => s,n,set(__tzone8=00:00-23:59)
exten => s,n,set(__tzone9=00:00-23:59)
;
exten => s,n,set(_dest1=01234123456)  ;dest1 emea_pager
exten => s,n,set(_dest2=001765412345) ;dest2 na_pager
exten => s,n,set(_dest3=006165453457) ;dest3 au_pager
```

```
exten => s,n,set(_dest4=08441231234) ;dest4 uk_no
exten => s,n,set(_dest5=001744519651) ;dest5 na_no
exten => s,n,set(_dest6=006118954654) ;dest6 au_no
;
exten => s,n,set(dialpre=9) ;dialing prefix
;
exten => s,n,set(dialcon=international) ;dialing context
;
exten => s,n,Goto(ftstimeing,s,1)
;
[ftstimeing]
;
;This sections runs though the days of the week and checks the time
;against DOW and time
;
exten => s,1,GotoIfTime(${tzone1}|mon|*|*?dest1,1)
exten => s,n,GotoIfTime(${tzone2}|mon|*|*?dest4,1)
exten => s,n,GotoIfTime(${tzone3}|mon|*|*?dest5,1)
exten => s,n,GotoIfTime(${tzone4}|mon|*|*?dest5,1)
exten => s,n,GotoIfTime(${tzone1}|tue|*|*?dest1,1)
exten => s,n,GotoIfTime(${tzone2}|tue|*|*?dest4,1)
exten => s,n,GotoIfTime(${tzone3}|tue|*|*?dest5,1)
exten => s,n,GotoIfTime(${tzone4}|tue|*|*?dest5,1)
exten => s,n,GotoIfTime(${tzone1}|wed|*|*?dest1,1)
exten => s,n,GotoIfTime(${tzone2}|wed|*|*?dest4,1)
exten => s,n,GotoIfTime(${tzone3}|wed|*|*?dest5,1)
exten => s,n,GotoIfTime(${tzone4}|wed|*|*?dest5,1)
exten => s,n,GotoIfTime(${tzone1}|thu|*|*?dest1,1)
exten => s,n,GotoIfTime(${tzone2}|thu|*|*?dest4,1)
exten => s,n,GotoIfTime(${tzone3}|thu|*|*?dest5,1)
exten => s,n,GotoIfTime(${tzone4}|thu|*|*?dest5,1)
exten => s,n,GotoIfTime(${tzone1}|fri|*|*?dest1,1)
exten => s,n,GotoIfTime(${tzone2}|fri|*|*?dest4,1)
exten => s,n,GotoIfTime(${tzone3}|fri|*|*?dest5,1)
exten => s,n,GotoIfTime(${tzone4}|fri|*|*?dest5,1)
exten => s,n,GotoIfTime(${tzone5}|sat|*|*?dest1,1)
exten => s,n,GotoIfTime(${tzone6}|sun|*|*?dest1,1)
;
;Fall through point
exten => s,n,Goto(dest1,1)
;
;Dialed using the Local channel so call handling is observed
;
exten => dest1,1,Noop(Calling ${dest1})
```

```
exten => dest1,n,Dial(Local/${dialpre}${dest1}@${dialcon})
exten => dest1,n,Hangup()

exten => dest2,1,Noop(Calling ${dest2})
exten => dest2,n,Dial(Local/${dialpre}${dest2}@${dialcon})
exten => dest2,n,Hangup()

exten => dest3,1,Noop(Calling ${dest3})
exten => dest3,n,Dial(Local/${dialpre}${dest3}@${dialcon})
exten => dest3,n,Hangup()

exten => dest4,1,Noop(Calling ${dest4})
exten => dest4,n,Dial(Local/${dialpre}${dest4}@${dialcon})
exten => dest4,n,Hangup()

exten => dest5,1,Noop(Calling ${dest5})
exten => dest5,n,Dial(Local/${dialpre}${dest5}@${dialcon})
exten => dest5,n,Hangup()

exten => dest6,1,Noop(Calling ${dest6})
exten => dest6,n,Dial(Local/${dialpre}${dest6}@${dialcon})
exten => dest6,n,Hangup()

exten => i,1,Hangup()
exten => t,1,Hangup()
exten => h,1,Hangup()
```

This can be expanded to include public holidays, if required. It can be possible to handle many years' public holidays in one line. For example, between the years 2009 and 2016, the UK's summer public holiday falls on the dates between the 25th and 31st of August and is always a Monday. Therefore, we have something like this:

```
GotoIfTime(*,Mon,25-31,Aug?dest1,1)
```

This will catch all UK summer public holidays, and as there are no other Mondays in August clashing with these dates, it's a set-and-forget for many years (just don't forget to change it after 2016!). The same goes for the majority of other public holidays except for Easter.

For these variable dates, we can resort back to the internal database to store the details and then use the `GotoIf()` application to check if the date is a holiday.

Variables

Variables are key to making the dialplan and system work in a manner that a user expects. The user would expect the system to know everything they have set on their extension, and not have to enter codes or dial special access numbers.

There are a number of places in which variables can be stored including the dialplan, `sip.conf`, `iax.conf`, `chan_DAHDI.conf` (in version1.6), and the Asterisk database (AstDB). For example, if we have a number of static dial strings we wish to store for each type of call and carrier we use, and then use them in a number of sections, the `[globals]` section of the `extensions.conf` file is the obvious place to declare them. If we wish to set a variable when a call is initiated from a SIP device, external caller ID or account codes are a good example, the `setvar` command in the `sip.conf` file is ideal for that purpose. Just remember that it won't work for calls sent to that device just when the calls are made. Finally, the AstDB is great for variables that are more transient in nature, such as call counts.

Inheritance of channel variables through the dialplan

On occasion, when using complicated dialplans you may wish for a variable's value to be kept as the call progresses. This is achieved by adding a _ [underscore] or a __ [double underscore] before the variable name.

A single _ will cause that variable to be inherited into the channel that started from the original channel, for example:

```
Set(_name1=value1)
```

If you want the variable to be inherited to all child channels indefinitely, then add __ before the variable name. For example:

```
Set(__name2=value2)
```

This should not be confused with setting the variable with the g option, as this sets it as a global variable. Doing so makes the variable available to all channels globally.

So, you may ask "why might we store dial strings as a variable?" The simple reason is that it allows a minimal amount of code for dialing all numbers, but still allows for different **classes of restriction**, by which we mean allowing different users to have different restrictions in what they can and cannot dial.

To pass these variables we will use a macro. **Macros** are like a template that we can use for repeated tasks, and they allow the passing of variables in an ordered fashion to the macro context. The call will jump to the s extension. The calling extension, context, and priority are stored in ${MACRO_EXTEN}, ${MACRO_CONTEXT}, and ${MACRO_PRIORITY} respectively. Arguments passed are accessed as ${ARG1}, ${ARG2}, and so on within the Macro. While a Macro is being executed, it becomes the context, so you must be able to handle the h, i, and t extensions if required within that context.

Let's build our small macro dialplan. We have a variable defined in the `globals` section of the `extensions.conf` file as follows:

```
[globals]
INT_CALL=IAX2/username@peer_out/
INT_CALL_ID=01234123456 ; default international callerID
INT_CALL_LIMIT=5 ; Limit on the number of calls
```

In the context that we use for dialing, we have:

```
; International long distance through trunk
exten => _90.,1,Macro(outdial,${INT_CALL})
```

Here, we have defined the macro we are going to pass the call to, along with a single variable we defined in the `globals` section (the value of the calling extension can be retrieved within the macro by using `${MACRO_EXTEN}`).

The macro context looks like this:

```
[macro-outdial]
exten => s,4,Dial(${ARG1}${MACRO_EXTEN:1},180)
```

This is the same as the dial string:

```
exten => s,4,Dial(IAX2/username@peer_out/01234123456,180)
```

We have seen that we can pass one dial string, but let's now pass other variables to the `Dial()` application, such as a backup route for outgoing calls, and the caller ID we want to use for the call.

```
exten => _90.,1,Macro(outdial,${INTCALL},${INT_CALL_ID},${INT_CALL_
LIMIT})
  [macro-outdial]
exten => s,1,Set(GROUP()=OUTBOUND_GROUP) ;Set Group
exten => s,2,GotoIf($[${GROUP_COUNT(OUTBOUND_GROUP)} > ${ARG3}]?103)
;Exceeded?
exten => s,3,Set(CALLERID(num)=${ARG2})
exten => s,4,Dial(${ARG1}${MACRO_EXTEN:1},180)
```

Now it's time to bring some `.conf` file variables into the mix. Using the `setvar` facility in the `sip.conf`, `iax.conf` and `chan_dahdi.conf` files, we can set variables specific for every user such as unique caller ID, call limits, whether we want to record the call, account codes. Basically, anything that will help you handle calls more efficiently.

```
setvar=account_code=2206
setvar=callidnum=01234123456
setvar=tenantID=2
```

One problem using .conf files is that the relevant channel module needs to be reloaded after a change, and in the case of DAHDI, Asterisk would need to be restarted. This isn't too much of an issue but the need can be removed by using the AstDB for storing commonly changed settings, such as caller ID and recordings.

You may think that all this variable use is over-complicated, but consider a system that supports multiple tenants. Using these techniques, you will only need one dialplan for multiple tenants instead of one per tenant. Simply set the tenantID in the relevant .conf file and then store the tenants' features in the globals section of the dialplan and in the AstDB, and all calls will go out as that tenant group. The concept is the same for other scenarios, such as departments that require cross charging of telephone costs.

Using the AstDB

Setting and retrieving variables in the AstDB is very simple and achieved through the use of the Set() application. Variables can exist in splendid isolation or be grouped into families. The syntax for setting a variable is:

```
Set(DB(family/variable)=value)
```

Retrieving the variable's value is equally as simple:

```
Set(result=${DB(family/variable)})
```

So, let's have a look at how we can implement a simple multi-tenant dialplan using multiple variable stores:

```
INT_CALL1=IAX2/username@peer_out_1/
INT_CALL2=IAX2/username@peer_out_1/
INT_CALL_LIMIT1=5 ; Limit on the number of calls
INT_CALL_LIMIT2=5 ; Limit on the number of calls
exten => _90[1-2]XXXXXXXXX,1,Set(INTCALL=INTCALL${tenantID})
exten => _90[1-2]XXXXXXXXX,n,Set(INT_CALL_LIMIT=INT_CALL_
LIMIT${tenantID})
exten => _90[1-2]XXXXXXXXX,n,Macro(outdial,${INTCALL},
${callidnum},${INT_CALL_LIMIT})
```

As we can see, we have been able to cut down the amount of code and make it universal for different types of users and systems. Using a macro lets us pass an ordered list of arguments. It is easiest to think of macro arguments as a list of variables since they are handled the same way.

Due to the way macro is implemented, it executes the priorities contained within it via a sub-engine, and a fixed per-thread memory stack allowance. Macros are limited to seven levels of nesting. It can be possible that stack-intensive applications in deeply-nested macros could cause Asterisk to crash. Take this into account and be very careful when nesting macros.

Dialplan features and additions

In this section, we are going to look at the DEVSTATE() function and the System() application. We will see how we can check and change the "status" of devices with the DEVSTATE() function and use the system application to cause scripts on the server to be run.

func_devstate

The func_devstate application allows the status of a peer to be known before you dial it. This is very useful in many applications. We will cover a few of them here but you will be able to find many more.

The func_devstate application is part of Asterisk 1.6, but Russell Bryant (of Digium) has a back-ported version for Asterisk 1.4. This can now be found at:

```
http://svn.digium.com/community/russell/asterisk-1.4/func_devstate-
1.4/func_devstate.c
```

For most Linux distributions, installing the function is pretty simple:

```
cd /usr/src/asterisk/funcs
wget http://svn.digium.com/community/russell/asterisk-1.4/func_devstate-
1.4/func_devstate.c
cd ..
make clean
./configure
make menuselect

Choose option  -> 6.  Dialplan Functions
Then make sure that you have an entry like 8.func_devstate
make
make install
```

If under `Dialplan Functions`, the `DEVSTATE()` function does not show up, you will need to edit the `menuselect-tree` to add it.

`<member name="func_devstate" displayname="Gets or sets a device state in the dialplan" remove_on_change="funcs/func_devstate.o funcs/func_devstate.so">`

Then compile Asterisk as shown previously.

What can we use the DEVSTATE() function for?

The `DEVSTATE()` function is versatile, allowing us to check and/or set the status of a device, as its name suggests. One very common use is to activate phone lamps, showing users if they have set a feature such as DND or call forwarding. In the following examples, we will look at both setting and checking methods:

The function reports on, or can set, the following states:

```
NOT_INUSE
INUSE
BUSY
INVALID
UNAVAILABLE
RINGING
RINGINUSE
ONHOLD
```

Outgoing trunk selection

The application can be used here to check that an outgoing peer is "available" and not "down", before you send a call to it. This is useful if you have peers or remote systems that are on variable quality connections.

```
exten => _90.,1,Macro(outdial,${PRIDIAL},${INT_CALL_ID},${INT_CALL_LIM
IT},${BAKDIAL},${PRIPEER})
[macro-outdial]
exten => s,1, Set(GROUP()=OUTBOUND_GROUP) ;Set Group
exten => s,2,GotoIf($["${DEVSTATE(${ARG5})}"="UNAVALIABLE"]?s,7)
exten => s,3, GotoIf($[${GROUP_COUNT(OUTBOUND_GROUP)} > ${ARG3}]?103)
exten => s,4, Set(CALLERID(num)=${ARG2})
exten => s,5, Dial(${ARG1}${MACRO_EXTEN:1},180)
exten => s,6, Hangup()
exten => s,7, Dial(${ARG4}${MACRO_EXTEN:1},180)
exten => s,8, Hangup()
exten => s,n, Noop(call Limit exceeded)
```

This example is an expansion of our previously used macro and has a couple of extra arguments passed to it. This makes it very flexible, as the backup peer can be different for each dialed number.

Calling extensions

The application lets you see if extensions are busy or "out of service" before calling them. This can be useful for handsets that support call waiting, but you don't want to fully disable it for all calls. Before calling the extension, you can check to see if the extension has call waiting enabled and then, depending on the result, check the device status as follows:

```
exten => 2XXX,1,Macro(dialext)

[Macro-dialext]
exten => s,1,NoOp(SIP/${MACRO_EXTEN} has state ${DEVSTATE(SIP/$
{MACRO_EXTEN})})
exten => s,n,Set(CW=${DB(CW/${MACRO_EXTEN})})
exten => s,n,GotoIf($["${CW}"="YES"]?dial)
exten => s,n,GotoIf($["${DEVSTATE(SIP/${MACRO_EXTEN})}"!="NOT_INUSE"]?
s-BUSY,1)
exten => s,n(dial),Dial(SIP/${MACRO_EXTEN},35)
exten => s,n,Goto(s-BUSY,1)
exten => s-BUSY,1,Voicemail(${MACRO_EXTEN},b)
exten => s-BUSY,n,Hangup()
```

In the previous example, we have used the internal database to set the flag to say if call waiting is enabled or not. If call waiting is anything other than YES, the status of the extension will be checked, otherwise the DEVSTATE isn't checked and the extension is just called. As we will see next, we can expand this to light a BLF (Busy Lamp Field) key as well, to give a visual indication to users of the device status.

Setting lights

We can also use the DEVSTATE() function to set BLF lights on and off, a very simple but highly effective feature. This is particularly helpful if you are using the dialplan for setting call forwards or DND. It can also show if a call center agent is logged in or not, on their phone.

To illustrate this functionality, we have a very simple example showing how to turn the light on and off. It uses one number to toggle the light status and is not specific for the particular phone—all phones dial the same number and it is the CHANNEL variable, which is used to set it for a specific phone. In this example, we have two hints—4078 and 4071, and these are linked to extensions 5078 and 5071.

Using this code and adding additional code to set the database key for call waiting (as we have already covered) would give the phone user a visual indication as to whether call waiting is set or not.

```
exten => 4071,hint,Custom:light5071
exten => 4078,hint,Custom:light5078
exten => 1236,1,Goto(1236-${DEVSTATE(Custom:light${CHANNEL:4:4})}|1)
exten => 1236-UNKNOWN,1,Goto(1236-NOT_INUSE|1)
exten => 1236-NOT_INUSE,1,Noop(Turn ${CHANNEL:4:4} light on)
exten => 1236-NOT_INUSE,n,Set(DEVSTATE(Custom:light${CHANNEL:4:4})=
INUSE)
exten => 1236-INUSE,1,Noop(Turn ${CHANNEL:4:4} light off)
exten => 1236-INUSE,n,Set(DEVSTATE(Custom:light{CHANNEL:4:4})=
NOT_INUSE)
```

By using `userevent`, you can also send out manager events to update the **Flash Operator Panel**. The following would set the CW flag for the **Flash Operator Panel** for our extension and change the icon to reflect the status.

```
exten => 1236-NOT_INUSE,2,UserEvent(ASTDB|Channel: ${CHANNEL}^Family:
CW^Value: SET ^)
```

> There is also a version of DEVSTATE() called EXTSTATE(). It is a modified version of the DEVSTATE() function that returns the state of an extension, rather than the state of a device. This means you can write dialplan logic based on the state of an extension (in use, ringing, on hold, and so on). The extension just needs to have a hint so we can determine which devices to check.

Boosting outgoing call capacity

We're going to have a look at how DEVSTATE() has been used to address an unusual situation. A call-centre customer wished to temporarily increase their outgoing call capacity, in this case by 20 concurrent calls, to cater for a particular project. However, in their location, with their budget and given the temporary need for extra capacity, the only effective means of boosting bandwidth is to utilize multiple ADSL circuits. In other words, SDSL and leased line circuits were too costly for consideration. Therefore, there was a need to **bond** multiple ADSL circuits together within Asterisk, in order to provide a single high-bandwidth circuit for outbound calls.

It may seem obvious, but when calculating call capacity with ADSL circuits, the figure we're interested in is the lower of the upload/download speeds. It doesn't matter if you have a superfast 20 MB DSL circuit, chances are that you only have an uplink speed of 800 kbps or less. It must also be remembered that once traffic exceeds 50% of the link speed, collisions and latency are likely to become an issue and must be addressed. This gives you a theoretical limit of up to 10 uncompressed calls per circuit if you're really lucky. Of course, with the GSM codec you can get a lot more, but at the cost of audio quality, which your customer is unlikely to accept. They expect PSTN quality and nothing less. If bandwidth utilization is on the borderline with an uncompressed codec, it can be advantageous to use a commercial (non-free) codec such as G729, which is obtainable from Digium.

Using multiple broadband lines

Using the criteria already discussed and assuming a 500 kbps uplink speed, it was determined that four broadband circuits were needed. This may sound expensive, but in reality it's not, when you consider that the alternative was 20 channels of a PRI (ISDN 30), which worked out at twice the cost of four PSTN lines with broadband. As we'll see later in the sales appendix, a major benefit of VoIP is that the customer is paying much less for line rental. This solution only reinforces that benefit.

We are going to describe a solution that used four broadband circuits, but another advantage of this approach is that it is very scalable. To illustrate, a system has been set up for a charity in the UK that had 75 agents placing thousands of calls a day on just eight broadband circuits.

 This example was tested and proven using Asterisk 1.4.19.2, and should work with releases up to 1.4.19.2 - it's operation cannot be guaranteed in other versions.

Configuration overview

Once the circuits are delivered, you will end up with four routers connected to the broadband service of your choice. Each router will have a unique IP address. In our case, we shall assume they are as follows:

```
192.168.1.1
192.168.1.2
192.168.1.3
192.168.1.4
```

We will also assume that the VoIP ITSP has multiple IP addresses that you can connect to, though if not, you can probably do some clever address translation in the routers.

Setting up the routing in Linux

Let's assume our VoIP ISP has provided us with the following external IP addresses:

```
88.88.88.81
88.88.88.82
88.88.88.83
88.88.88.84
```

Within Linux, you can easily set up different gateway addresses for a given destination. The file that manages the gateways is normally called `ifup-routes` in the `/etc/sysconfig/network-scripts` directory.

To configure the gateways, we append the following to the `ifip-routes` file:

`/sbin/route add 88.88.88.81 gw 192.168.1.1`

`/sbin/route add 88.88.88.82 gw 192.168.1.2`

`/sbin/route add 88.88.88.83 gw 192.168.1.3`

`/sbin/route add 88.88.88.84 gw 192.168.1.4`

Taking the last entry, what we're saying is that for all traffic to `88.88.88.84`, route it via the router at `192.168.1.4`.

If you reboot and run the command `route -n` in a terminal session, you'll see these routes in place.

Configuring Asterisk

We now turn our attention to the Asterisk configuration. When we make a call, we're going to keep count of how many calls we have on a broadband line, so that when the circuit is "full", we can move on to the next available one.

Firstly, in the `extensions.conf` file, we need to declare a variable that sets the maximum concurrent calls we will allow through any one router.

```
MAXVOIPCALLS=5          ; Maximum Calls we allow over IP (outbound)
```

We've previously set up four entries in the `iax.conf` file called `iaxline1` through to `iaxline4`. They have identical entries, with the exception on the `host=` line. Here we assign the appropriate external IP address, that is, `88.88.88.81` for `iaxline1` and so on.

Now, we need to declare the IAX channels as follows:

```
AIXVOIPOUT = IAX2/iaxline1
AIXVOIPOUT1 = IAX2/iaxline2
```

As expected, we use a macro to manage the call routing. The example below only shows two lines for the sake of brevity.

```
[macro-voiptrunk]
exten => s,1,Noop(Number of Broadband calls)
;Use devstate to test the availability of the trunks. You could put
some code in, to use an alternative if they are off line
exten => s,2,Noop(Trunk 1 ${DEVSTATE(${AIXVOIPOUT})} -> ${GROUP_
COUNT(VOIPTRUNKS@list1)})
exten => s,3,Noop(Trunk 2 ${DEVSTATE(${AIXVOIPOUT1})} -> ${GROUP_
COUNT(VOIPTRUNKS@list2)})
exten => s,4,GoToIf($[${GROUP_COUNT(VOIPTRUNKS@list1)} <
${MAXVOIPCALLS}]?10:20)
;Have we exceeded the max calls per trunk?  If so, jump to Extension
20 and use second trunk
exten => s,10,gotoif($[${DEVSTATE(${AIXVOIPOUT})} =
UNAVAILABLE]?20:11)
;Here we test to see if the trunk is available, if it's gone off-line,
we use the second trunk
exten => s,11,Set(GROUP(list1)=VOIPTRUNKS) ;increment the usage count.
exten => s,12,noop(${DEVSTATE(${AIXVOIPOUT})})
exten => s,13,Noop(Trunk 1-> ${GROUP_COUNT(VOIPTRUNKS@list1)})
exten => s,14,Dial(${AIXVOIPOUT}/${MACRO_EXTEN})
exten => s,15,Goto(s-${DIALSTATUS},1)

exten => s,20,GoToIf($[${GROUP_COUNT(VOIPTRUNKS@list2)} <
${MAXVOIPCALLS}]?21:40)
exten => s,21,Set(GROUP(list2)=VOIPTRUNKS)
exten => s,22,Noop(Number of Broadband calls)
exten => s,23,Noop(Trunk 2-> ${GROUP_COUNT(VOIPTRUNKS@list2)})
exten => s,24,Dial(${AIXVOIPOUT1}/${MACRO_EXTEN})
exten => s,25,Goto(s-${DIALSTATUS},1)

exten => s,40,Congestion(15) ; No more lines left
```

Explanation of the macro

Priorities "2" and "3" use the DEVSTATE() function to test the availability of the broadband lines. If a line is down, UNAVAILABLE will be returned. At "4", we look to see if we've exceeded max calls on this line. If we haven't, we'll place a call on the first router, otherwise go to the next. "11" records the in-use count and increments it for the given router (list1).

What happens, overall, is that the first 5 calls (set by the global MAXVOIPCALLS) will go via router 1, the sixth will go via router 2. If in the meantime a call is dropped from router 1, the next call placed will go back to router 1, even if other calls are ongoing on router 2.

Finally, we need to add a call to the macro in our dialplan:

```
[outbound-national]
exten => _0Z.,1,NoOp(national call)
exten => _0Z.,2,Macro(voiptrunk)
```

The above technique is scalable. You can add as many broadband lines as you need. The end result is that you can say to your customer, "want more outgoing capacity? Just add another DSL line". However, it must not be forgotten that there may be more stable solutions such as SDSL and leased lines, depending on location.

Downsides

The above example works really well for outbound calling, but not so well for inbound. If you own the server the customer is connecting to, then you can reverse the logic at your end. If you don't, then all you can do is allocate one router for inbound (register via an inbound router) and the rest for outbound.

System() application

The System() application allows Asterisk to run Linux commands and shell scripts. What we will look at here is a simple **hotdesking** deployment script for asterisk. This type of deployment method is used by all commercial PBXs and is needed for any enterprise deployment of Asterisk. Hand editing filenames or even configuring phones via their web GUI will not be accepted by a customer or end user.

The dialplan is very simple. The user dials a code from his/her handset and is asked to enter a four-digit number (their hotdesk ID). The dialplan then stores this as a variable. It also sets the caller ID and the IP address for the set and passes these to the script.

```
[hotdesk_in]
exten => s,1,Answer
exten => s,n,Playback(privacy-thankyou)
exten => s,n,Read(MY_EXTEN,access-code,4)
exten => h,1,Set(MY_IP=${SIPPEER(${CALLERID(num)}:ip)})
exten => h,2,system(/usr/local/sbin/exten_in ${MY_EXTEN}
${CALLERID(num)} ${MY_IP})
```

With these three variables, the script then knows the handset's existing number, IP address, and the number it wants to be.

In addition, the script then performs an **Address Resolution Protocol (ARP)** lookup on the IP address to find the phone's MAC address. It needs this because, as in the example, we are using the phones config file in the format of <MAC-ADDRESS>.cfg, and we configure the sets via **TFTP (Trivial File Transfer Protocol)**. Hence, as we know the MAC address we can copy the config files to the correct name.

Firstly, we will copy the old <MAC-ADDRESS>.cfg to a different file name. Then we copy the config file for the extension number we wish the phone to be using the MY_EXTEN variable we have passed to the script to define it to our new <MAC-ADDRESS>.cfg. Now when the set reboots, it will pick up the new file. However, we want this to be automatic and with as many handsets as it can have. The sip notify command does so when configured in the sip_notify.conf file, in the case of Aastra handsets, as follows:

```
[aastra-check-cfg]
Event=>check-sync
Content-Length=>0
```

The following command will cause the phone to check the config file for changes and reboot if any change is found:

/usr/sbin/asterisk -rx "sip notify aastra-check-cfg ${CALLERID}"

When it reboots, it will pick up its new configuration. By using scripts such as the previous one, you can speed up "moves, adds, and changes" and cut out the need for engineers to put out or replace handsets. It can also be used to provide a form of hotdesking with the user dialing a code to set the handset as theirs, and then log out when they leave (copying back the previous config), thus returning the phone to its previous state.

Summary

In summary, we have looked at how to break down your dialplan into small, manageable contexts or objects. These can then be included into the dialplan to create a system with the flexibility to match any commercial PBX. We also looked at improving the security of the dialplan such that it is easy to manage who can dial where in an understandable way.

We looked at the many different ways that variables can be stored in the system and called upon when required, as well as seeing how they can interact with macros to make the dialplan more streamlined. Here we used one macro for many extensions.

We looked at the DEVSTATE() function and the uses that it can be put to. These are not just (as it initially seems) for checking the status, but also a way to set the status and light a BLF key to show a feature is set.

We looked at time and day call routing, and how it can be used to route calls based on time and day. We also looked at the clever use of date ranges, so that we can future-proof our dialplans for holidays for many years to come.

And finally we looked at the System() application and how this can be used for easing the deployment of handsets in an enterprise solution. In the next chapter, we shall focus on exploring network considerations.

2
Network Considerations when Implementing Asterisk

The reasons for choosing a Voice over IP (VoIP) telephone system in preference to a traditional PBX are many, as evidenced by the growth of IP-capable PBXs in larger businesses for some years now. One of the reasons is the opportunity it affords a business to maximize the investment it has made in its IT infrastructure. Rather than create and maintain one mechanism for transporting computer communications traffic, and another for transporting voice communications traffic, a company can simplify matters by carrying all communication traffic on a single, IP-based network.

However, putting all such traffic through one channel means that it is more important than ever to ensure it is up to the task. The irony is that, should you have to choose between an IP data network and an analog voice network based on stability, then the likely choice would be the voice channel. After all, most people have higher expectations of telephone infrastructure than of computer networks—they expect the phone to work, every time. Therefore, over time, pure voice communication transport mechanisms (initially analog but for some time now digital in nature) have developed with the emphasis very much on stability and availability above all else. This has meant that simplicity is preferable, and change is slow to occur. Of course, when designing a network with stability as the prime goal, having only one type of traffic to carry helps enormously.

But the voice channel is not suitable for network traffic, unless you hanker for a return to modems on every desk. The IP-based Ethernet network, on the other hand, is now eminently capable of being used for voice traffic. Indeed, some of the features of IP networks, such as the ability to choose between multiple routes on the fly, introduce elements of redundancy that have previously been difficult and/or expensive to add to voice networks.

In this chapter, we will look at the considerations you need to bear in mind when introducing a Voice over IP system to a company. We will discover what makes the difference between giving your customer a telephone setup that rivals a good circuit switched system for call clarity and uptime, while improving the features and capabilities many fold. After all, most people will not judge your shiny new phone system a success if they cannot simply pick up the phone, dial a number, and have a conversation.

Centralized and distributed installations

Let's start with a brief definition of **centralized** and **distributed installations**. A centralized installation (sometimes called a "System" installation) is where there is a physical installation of Asterisk onsite, and by that we mean connections from the phone are on a high speed network (Cat5 and so on) where bandwidth and latency should not be an issue. A distributed, or hosted installation, is where the end user is using a remote server to handle the PBX functions. Typically, such installations have a small number of IP phones/adapters onsite, for instance where a small remote office is required to connect to the company's centralized PBX installation. This term is also used to describe situations where the customer doesn't have a PBX of their own, and rents PBX services from someone else.

Centralized installations

With the local PBX setup, which is a centralized installation, unless you have major problems with your LAN there should be no issue with communication between endpoints, such as handsets, and the PBX. If you find that handsets are losing connectivity, then you need to start investigating why this is the case before you proceed any further with implementing or enhancing the system. Otherwise inefficiency of the network is going to cause you a multitude of issues further down the line.

Distributed solutions

We're assuming here that you're going to be running an Asterisk installation at some central site and servicing the remote phone's requests. This may be a billable service you are supplying or a centralized system at your customer's premises that is servicing multiple locations.

With such a setup, a major concern is bandwidth, particularly if you have quite a large number of extensions on a site with no PBX. The reason for this is simple, each call in a hosted environment, whether internal or external, will be required to traverse the WAN to the central PBX. Indeed, internal calls in a remote office are a double whammy, as they are routed via the WAN to the PBX, and then routed back to the same office! As a rule of thumb, it is worth seriously considering moving away from a hosted setup if an office contains more than 50 or so extensions. Although this can come down significantly if available bandwidth is quite limited or internal calls are higher than one would normally expect for an office of that size. The good news is that, for a smaller office, the hardware specification required to run an Asterisk server capable of handling the lower call volumes is modest indeed.

Latency and jitter

With any telephony system, bandwidth will determine the maximum number of concurrent internal and external calls that can be made. However, bandwidth is far from being the only consideration, irrespective of whether your installation is centralized or distributed. The latency characteristics of the circuits, LAN, and WAN, will dictate how well those calls get carried, and how the quality of those calls is perceived.

 Latency is the amount of time a message takes to traverse a system.

There are many factors that can influence latency including processor speed, available memory, disk spindle speed, and so on. However, in virtually all hosted installations communication circuits is the defining factor. When there is a limited amount of latency that a commercial system can bear, choosing the most appropriate circuit provider is vital. In fact, choosing a circuit with low latency is arguably even more important than reliability, as it is easier to introduce redundancy to communications circuits based on availability (if circuit A is down switch to circuit B) than it is based on latency (if circuit A is showing high latency switch to circuit B).

Asterisk as a PBX also copes much better than phones with high latency, and has mechanisms to help, such as the `qualify=` and `qualifysmoothing=` entries in `sip.conf` and `iax.conf`, which ensure that endpoints are pinged by the server periodically to check that they are still available.

 In larger organizations, care should be taken with the use of `qualify=yes` as it tends to ping all the endpoints at the same time, generating a temporary packet "storm". If endpoints de-registering is an ongoing issue, and your endpoints have the facility to generate their own registration "keep-alive" traffic, then that is a better solution.

As a result of these measures you should only have audio delay and/or echo to worry about, and the latency mentioned above will be the likely cause of that.

You can perform a quick audible echo check by using the Echo() dialplan application in Asterisk like this:

```
exten => *77,1,Echo()
```

This will allow anyone that dials *77 to hear any echo, as their audio is fed straight back to them.

It is also worth mentioning jitter in relation to circuit quality. Whereas latency is the delay between a message being sent from one side of a circuit and received at the other, jitter refers to how consistent this delay is.

Jitter is the variation in the time between packets arriving, and can be caused by network congestion or changes to the route traversed by concurrent packets.

So while you can measure latency in terms of a single packet traversing a circuit, you need to measure jitter over time, to see just how consistent that delay is. A low-jitter circuit will tend to deliver packets in pretty much the same order they are transmitted. If jitter is high then packets will start to arrive out of order, with the result that the audio will start to distort.

To illustrate latency and jitter, have a look at the sample graph that follows:

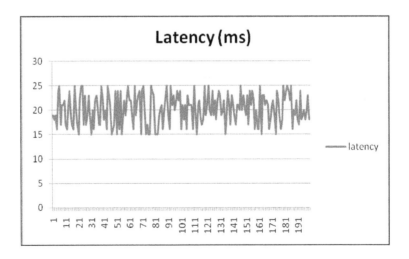

As you can see, the latency over time is consistently between 15 and 25 ms, which for a WAN is not at all excessive. However, it is far from being a smooth graph, and in real life such "jitter" on a circuit would result in a significant deterioration in the quality of the transmitted audio as many packets would be delivered out of order to the endpoint.

Bearing latency and jitter across a WAN in mind, many companies choose to forego the opportunity to save a few pennies by ignoring domestic-grade circuit options (predominately Asymmetric Digital Subscriber Line [ADSL]). The major commercial circuit providers in the UK, such as BT, Tiscali, and the likes are capable of providing circuit SLAs that not just guarantee a certain level of availability but also provide guarantees about latency. As always, it is worth doing your homework before committing to a choice that can make such a huge difference to end-user perception of the system.

There is an Asterisk dialplan called `MilliWatt()` which, when run, produces a continuous tone of 1004 Hz:

```
exten => *78,1,MilliWatt()
```

When listening to this tone, distortions or breaks can be an indication that jitter is present and should be investigated. The likely cause will be the LAN/WAN infrastructure between the PBX and endpoint.

Unfortunately, in any call that terminates outside your company, you only have control over part of the picture. Once the call traffic leaves your system you are at the mercy of the Internet, and the other party's system. If they are still stuck in traditional analog telephony land, then latency and jitter at their end is likely to be low and any problems are probably yours. If they are enlightened VoIP users, then excessive latency or jitter may be down to their LAN/WAN, the Internet, or your LAN/WAN. So you need to make sure your house is in order before you start pointing fingers. However, if you have set up your system with low latency and jitter as a priority, and you run regular checks to make sure that those latency targets are being met, then you should be able to sleep easily. Indeed, having carried out the necessary investigations, research and tuning to ensure yours is a low-latency, low-jitter Asterisk installation, this could be an opportunity for you to share your knowledge, for an appropriate fee of course!

So how do you identify and address these issues? On an Asterisk server, you can get a good idea of the latency between hosts and peers by running the `cli` commands `iax2 show peers` or `sip show peers` — as long as you have `qualify=yes` in the peer profile. On the PBX CLI type:

```
iax2 show peers
```

Name/Username	Host		Mask	Port	Status
IAXTrunk1	217.14.138.130	(S)	255.255.255.255	4569	Unmonitored
201	192.168.10.201	(D)	255.255.255.255	4569	OK (4 ms)
IAXTrunk2	(Unspecified)	(S)	0.0.0.0	4569	Unmonitored

```
3 iax2 peers [1 online, 0 offline, 2 unmonitored]
```

`iax2 show peers` (above) lists a number of `iax2` trunks (status `Unmonitored`) and an `iax2` extension. The extension is showing a delay of 4 ms, which is indicative of an extension on the local LAN (which this is). On the PBX CLI type:

```
sip show peers
```

Name/username	Host	Dyn	Nat	ACL	Port	Status
SIPTrunk1	217.10.79.23		N		5060	OK (79 ms)
SIPTrunk1	217.10.79.23				5060	OK (79 ms)
251/251	192.168.10.201	D	N		5060	OK (54 ms)

`sip show peers` (shown above) also shows registrations of both trunks and extensions. In this case, we can see that the local extension 251 is showing a delay of 54 ms, which is high for a device on the same LAN and should be investigated.

> The SIP delay time is actually the time taken by the device to respond to a `notify` command rather than just a ping response, so can be affected by processing time on the device.

If all devices on the network support ICMP, and I would certainly recommend that you allow ICMP internally, then ping can also be used on the server to verify this information. If the ping delay is similar to the delay shown in the commands above, then you have a latency issue on your LAN/WAN. If the ping command is much lower, then the problem is likely to be with the device itself.

Tracking the cause of latency issues, particularly in a distributed environment, can be greatly assisted by the use of the `tracert` and `tracepath` commands. If a packet has to pass through a number of routers and switches to get from server to device, these commands will give you an indication of how long each step takes. It can even reveal that unexpected routes are being traversed. The commands are virtually synonymous, so use whichever you prefer.

Jitterbuffer

If you wish to look at the latency and jitter information for an active IAX2 (Inter-Asterisk Exchange) call then you can use the `iax2 show netstats` command, which shows the latency, jitter, and lost packets for all active IAX2 calls. This command assumes that you have enabled the Asterisk jitterbuffer on the IAX2 channel. The jitterbuffer is a mechanism for dealing with excessive jitter on a channel. Up to version 1.4 of Asterisk, the jitterbuffer only worked on IAX2 and ZAP channels, but from 1.4 onwards, it will also work with RTP channels such as SIP and H.323. The jitterbuffer works by buffering incoming packets, examining their timestamps and, where possible, re-ordering them so that they are delivered to the endpoint in the right sequence.

> Jitterbuffers work best as near to endpoints as possible. If an endpoint has its own jitterbuffer capability then that is usually preferable to it being carried out on the PBX.

To enable the Asterisk jitterbuffer for a SIP channel, for example, you should add the following lines to the `[General]` section of your `sip.conf` file:

jbenable = yes|no (Enables the use of a jitterbuffer on the receiving side of a SIP channel.)

jbforce = yes|no (Forces the use of a jitterbuffer on the receive side of a SIP channel. Defaults to "no".)

jbmaxsize = Number (Max length of the jitterbuffer in milliseconds.)

jbresyncthreshold = Number (Jump in the frame timestamps over which the jitterbuffer is resynchronized. Useful to improve the quality of the voice, with big jumps in/broken timestamps, usually sent from exotic devices and programs. Defaults to 1000.)

jbimpl = fixed|adaptive (Jitterbuffer implementation, used on the receiving side of a SIP channel. Two implementations are currently available - "fixed" (with size always equals to jbmaxsize) and "adaptive" (with variable size, actually the new jb of IAX2). Defaults to fixed.)

jblog = no|yes (Enables jitterbuffer frame logging. Defaults to "no".)

Care should be taken in using `jbforce`, as it will introduce a delay for all inbound traffic, whether it has excessive jitter or not. Indeed, for this reason, you should only consider the use of a jitterbuffer at all if you are finding that jitter is adversely affecting the quality of a significant percentage of calls. If your circuit latency is marginal, then adding a jitterbuffer into the mix could introduce enough latency to make echo detectable on calls.

Echo

These commands should provide you with enough information in most cases to at least give you a head start on the likely causes of any latency issues. Thereafter you can use commands such as `sip set debug` or `iax2 set debug` to gather more detailed information on the fly which can then be examined in the log.

So what level of latency in a system can be deemed to be acceptable? It's well documented that the human ear can detect delays in audio above 300 ms, so if there is latency greater than 300 ms, the user will hear an audible delay or even an echo. Why? Back in days of yore, telephone engineers found that they needed to feed what you were saying back into the ear piece, a feature known as sidetone. If they didn't, then the user would think they'd been disconnected. Of course, traditional PSTN systems, with their inherent low latency, only had an issue when large distances were involved. In other words, national calls tended to be fine, but international and intercontinental calls suffered to a greater or lesser degree from echo as well as significant delay between the transmission and reception of the voice signal. Bad echo makes it very difficult to concentrate on a conversation.

Another downside of high latency is that not all IP phones cope well with it, although some do better than others. For example, experience has shown Linksys phones to be good in distributed installations. Others will fail to register overtime and therefore go off-line. This isn't always immediately obvious, either leading to the potential for important calls to go straight to voicemail, or even be lost altogether.

Do your homework

Coming back to the customer's network, how do we ensure that it will carry the expected voice traffic without excessive latency or jitter, and with adequate availability? It's difficult to guarantee a high level of network availability to a customer without knowing what you have in the first place. While not usually being up to the standard of voice networks in terms of reliability and call quality, Ethernet-based IP network are so prevalent these days that many SMEs (Small and Medium Enterprises) follow the well-worn adage that *if it ain't broke then don't fix it*, even if the stability of the network is not ideal. Remember, though, that we are now planning to introduce a system that demands the highest levels of availability. The 99.5% uptime figure that the customer might have deemed perfectly adequate in the context of their data network, is nowhere near high enough for a telephone system, equating as it does to 18 seconds of silence in an average hour. Even worse is the fact that any active calls will be dropped unceremoniously.

Don't forget also that your Asterisk system is going to place new demands on the bandwidth of your customer's network, both internally and on the route(s) to the Internet. It is true to say that decisions on such things as the choice of codec used for the voice traffic will determine your overall bandwidth requirements, and it is quite typical for these decisions to be informed by the currently available bandwidth, and of course the cost of improving either its capacity and/or quality if required. The work to determine exactly where this improvement is needed can actually be quite an easy 'sell', particularly if the customer is experiencing difficult-to-trace network performance issues. It is not unusual to unearth, say, a domestic 10 Mbps hub hidden away under a desk, causing all sorts of network performance issues. Removing such obvious no-no's could probably reduce, maybe even completely negate, the need for a network upgrade.

Therefore, your first task should be to carry out an audit of the customer's network. The extent of this audit is reliant on such factors as the customer's existing network documentation, their budget for the new VoIP system, and their commitment to doing whatever is needed to ensure the success of the new system. The form of the audit can be anything from a perusal of their documentation, or a quick visual inspection (looking for those errant hubs), through to a complete 'sniff' of the network measuring traffic throughout. It is not unusual for more detailed audits to be outsourced to specialist IP Networking companies that have in-depth skills and the right equipment for the job. For this decision, there is a cost/benefit calculation to be made that will undoubtedly vary with each customer. However, it is perfectly possible to get adequate information that will allow you to gain a good understanding of the customer's network throughput without outsourcing the work. If the customer uses managed switches, then the job will be pretty easy, as virtually all will allow you to see real-time traffic information and download that data for later analysis. This is also true for the majority of routers, although that does rely on the customer having access to the router interface. If the router is managed externally then there will be a need to request the data from the service provider.

Should the customer not have a means of extracting the data from switches and/or routers, then you will need to consider the use of network monitoring software. In the open source world, ntop or, at a pinch, Wireshark (formerly Ethereal) can be used to monitor bandwidth usage. In the commercial sphere, companies such as Solarwinds and PacketTrap are just two that have established and very capable products.

While it's not essential to have the software and skills to carry out an effective audit of a customer's network, if you're planning to offer the "complete package" then it would be wise to develop your competency in this area to a point where you can deal with the simpler situations. After all, if you can give an immediate overview assessment of the quality of the network then you can only impress prospective customers. Once signed up, ongoing network monitoring would be a useful service to offer too, as it allows you to be more proactive. Most customers are impressed when you tell them they have an impending problem that they were not aware of themselves.

SLAs are for everyone

The audit should be aimed at providing the customer with a roadmap to improving their network up to agreed levels of quality, stability, and resilience. As already stated, voice networks have an availability and quality expectation that is far in excess of the typical SME computer network. It is not unusual for availability expectations to be five nines (99.999%), and for the supplier (yes, that's you) to provide a **Service Level Agreement (SLA)** with penalties if that figure is not achieved.

Some suppliers fight tooth and nail to reduce that SLA figure, but there is no reason to do so as long as you realize they can be a means to gain agreement on who is responsible for which part of the overall system up front, and might possibly differentiate you from the competition if you are in a commercial situation. As long as both you and your customer, whether they are internal or external, are being realistic, there is no reason why you cannot specify that certain preconditions are met before such an SLA can be offered, such as a certain level of bandwidth and latency across the network. You would also need to agree on the line of demarcation between "your" part of the system (such as the server, phones) and the customer's part (such as the underlying network) once any initial implementation work has finished and the enhanced network has been handed over to the customer's operations and maintenance team. But all such discussions should be approached with a view to the ultimate objective. After all, the goal for both sides should be to end up with a great phone system.

Achieving the goal

By now, you've probably agreed with the customer that their network needs work to achieve the agreed service levels. Let's have a look in a little more detail at how that network should look, bearing in mind all the time that the final product will be unique for each customer, given their different needs. Almost always, though, the emphasis will be on making improvements to achieve higher levels of quality, stability and resilience. Here we will look at what is required to achieve the required quality standards. Chapter 7 will explore the topics of stability and resilience in more detail.

In IP networks, quality is usually synonymous with dropped packets, or more accurately with the lack of dropped packets. Traffic across an IP network is not sent in a continuous stream, as in an analog voice network, but rather packaged up into discrete units and sent out into the network with a destination address as part of the package. As suggested by the terminology used, it is not dissimilar to sending packages by post, although in the network each package is pretty much the same size, so larger items are broken up before being sent and reassembled at the destination. It's a bit like getting your flat-pack furniture piece by piece.

Many things can cause these packets to go astray, but the IP protocol has a built-in mechanism (**TCP** or **Transmission Control Protocol**) for recognizing this has happened and requesting a resend of the package. This is achieved by including a checksum with each package to ensure that its data is received intact, and by acknowledgement of each package being sent back to the transmitting node. TCP is used for quality-sensitive data, that is, data that requires all the packages to be reassembled if it is to make sense. This is because there is no significant "storage area" for packets on a network, they travel at a certain speed and if they can't be dealt with inside a predefined time, then they are dropped.

For most types of traffic re-sending is usually very effective. After all if your 20 MB file download is delayed by a fraction of a second while a package is re-sent then you probably won't even notice. However, there is an appreciable amount of extra data and traffic generated, which can cause problems in networks that have a time-critical quality requirement, such as voice networks. Therefore, an alternative protocol (**UDP** or **User Datagram Protocol**) is used for voice and other "streamed" traffic, as there is no significant loss of perceived quality if small numbers of packets go astray between source and destination. The problems, such as latency and jitter, arise when larger amounts go missing in action.

Within company IP networks, dropped packets typically start occurring when traffic levels get too high. An analog voice network works by setting up a channel with more-than-adequate "bandwidth" between the two parties that is not used by anyone else. Think back to the old switchboard operator moving plugs around on a board, this is exactly what she was doing. An IP network, on the other hand, works on a similar basis to a sorting office, where everything posted converges on a central place (the router on a network) where the address is read and it is sent along the correct path towards its ultimate destination. However, if there are lots of letters and packages in the system, then post boxes can overflow, pickups can be missed, sorting equipment and people overloaded and the whole system would struggle to cope. So it is with a computer network.

Dropped packets in a computer network can frequently be traced to mismatched components, such as the aforementioned hub being used to allow a number of desks to utilize a single network floor port. It's all very well specifying that your comms cabinet be stuffed with Gigabit switches, and that all your PCs have 100 Mbps capability, but stick a 4-port 10 Mbps hub in the middle and the PCs plugged into it will be throttled to 2.5 Mbps or less. Then start adding IP phones on to each desk, or using softphone software on each PC, and you can see how it all gets messy very quickly.

At this point it is worth saying that, for a network carrying voice traffic, using network switches instead of hubs is a no-brainer. Switches go much of the way towards setting up dedicated channels between one port on the network and another, and are much more effective than hubs. The price difference between hubs and switches has reduced significantly in the recent past, removing the greatest barrier to their adoption. If your customer insists on an SLA, you should only agree if their network is fully switched. In fact, being realistic, you should only ever implement a VoIP solution over a fully-switched network as using hubs **will** cause you problems sooner or later by undermining any other steps you may take to ensure high quality voice traffic.

Furthermore, at the time of writing, Gigabit switches have reduced in price to the extent that serious consideration can be given to using them instead of 100 Mbps switches throughout the network. After all, many new PCs these days come with Gigabit network cards as standard, so putting 100 Mbps switches in your network will, again, artificially throttle speeds. Be aware, though, that there can be a significant difference between a cheaper Gigabit edge switch and a top-end Gigabit core switch, particularly in processing speed. In most cases it's better to have a good quality 100 Mbps core switch than a cheap Gigabit core.

It can be a big task to bring Gigabit ethernet to the desk, though, if the current wiring is not up to scratch. Hence, should the introduction of Gigabit throughout the network be out of scope then there is no serious need for concern. A well designed and implemented 100 Mbps network has plenty of bandwidth for voice traffic, even if it is not as future-proof as it might be.

Backups

One area that can cause problems is if data backups are run across the network during working hours. You'd be surprised how many companies run their backup routines at times when high call volumes can be expected, even during the lunch hour. Such activities have the potential of sucking up 99% of the available bandwidth. For this reason, it is not unusual for larger companies to route server backup traffic over a physically separate network to guarantee that it will not adversely affect front-office activities. For smaller companies this may be too costly an option, so a backup "window" needs to be identified that will occur during low call volume times.

To share or not to share

When designing your voice-capable network, one important consideration is whether or not to carry voice and data traffic on the same network infrastructure. In other words, should you implement a completely separate IP network for your phones, PBX, and Internet circuit? While there are significant benefits to doing so, particularly in terms of assuring bandwidth for voice traffic, in general it's probably not sensible for a couple of reasons.

The first reason covers practicality, particularly around the availability of floor or wall network ports. As many offices were designed around the desire to have one port per desk, it is usually a pretty disruptive exercise to increase that to two ports per desk, as you will need to do if you are to physically separate the voice and data traffic. The customer may also find the increased cost of switches and so on to be quite impractical. The "two ports per desk" impracticality can be assuaged somewhat through the use of VLANs (Virtual LANs), whereby logically separate LANs run side-by-side on the same physical equipment. Often VLANs are implemented by assigning switch ports into groups that allow them to communicate with each other but not with ports in other groups.

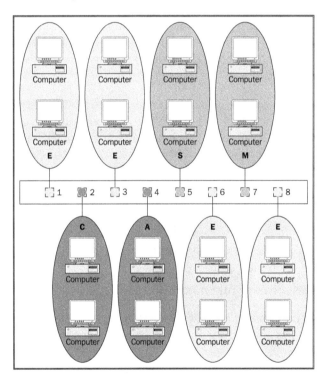

In the preceding diagram, we can see that PCs connected to ports 1, 3, 6, and 8 on the switch are on the same VLAN, thus they can communicate with each other but not with any of the PCs connected to other switch ports. Be careful, though, as port-based VLANs will ensure that each port switch will only carry one type of traffic (VoIP or data), which will not remove the need to have two ports per desk if complete traffic segmentation is a "must have" requirement.

There are alternative mechanisms for implementing VLANs, such as tagging traffic based on subnet, protocol, or MAC address. All serve the purpose of separating LAN traffic into logically separate networks on a single physical network, but as the segmentation is not based on a physical connector (that is, the switch port) they will allow for a desk's VoIP and data traffic to be routed over a single network port. Most business-grade IP phones now come with a two-port switch as standard, which allows you to plug the phone into the existing network port, and then plug the PC into the phone. A MAC, subnet, or protocol-based VLAN will then allow you to logically segment the VoIP traffic. The downside of this setup is that if the phone needs to be restarted, there will be a temporary interruption of the network connection to the PC. Make sure the customer is aware of this.

VLANs have the advantage of reserving bandwidth solely for voice traffic, while not necessarily needing to incur the extra cost of multiple switches. This does assume that you have an adequate number of ports free on your current switches if you decide to use port-based VLANs. However, the cost savings are greater when volumes are higher, as you can replace many smaller switches with fewer larger ones. As we will see later, using VLANs can also bring some advantage in increasing resilience to switch failure.

One interesting way around the issue of floor port availability is to use wireless technology for desk phones. Both Wi-Fi (802.11 in its various flavors) and DECT are gaining ground in the VoIP handset market, partly because they neatly sidestep this concern. They also allow for mobility within the workplace in situations where that is desirable, warehouses for instance. In a world where mobile telephony is firmly ensconced, they do not look out of place, although moving from desk phones to mobile handsets in some companies can still be too much of a culture change. It is fair to point out, too, that wireless technology has its downsides, such as:

- The cost of a wireless-based system will probably be greater than the one based on desk phones.
- Wi-Fi phones are currently notorious for having poor battery life.

The Wi-Fi or DECT network of repeaters needs careful design to ensure there are no poor signal areas in and around the buildings and to ensure that seamless handover from one base-station to another can occur. Be aware also that the choice of VoIP codec and Wi-Fi protocol can limit the number of simultaneous calls per base station.

[You can get more information on this at: `http://www.oreillynet.com/etel/blog/2005/06/maximum_number_of_voip_telepho.html`]

For now, DECT technology is mature enough to warrant serious consideration, but Wi-Fi solutions seem to retain too many issues still, although it is a rapidly maturing marketplace.

Another area that is often overlooked is power. Perhaps it's a hangover from analog phone days, but it's not unusual for customers to be unaware that an IP phone is a powered device. As it is frequently the case that there are too few power sockets for each desk in an office, use of **Power over Ethernet** (**PoE**) capable phones and switches is worthy of very serious consideration in all roll-outs. The added cost is only significant in systems with a small number of extensions, and even in these cases it may well be that only certain key phones need to be powered in the event of an outage, reducing the number of PoE switches required. By providing battery backup for the PoE switch and other critical devices, service can be maintained for short periods.

The other downside of completely separating voice and data network traffic is that **Computer Telephony Integration** (**CTI**) becomes more difficult to implement. Many IP phones these days have bundled with them software that allows for computer integration, such as selecting a contact from your address book and clicking a button on screen to initiate a call, or having a contact's details pop up on screen automatically when they call you. If, for instance, a company relies heavily on its CRM software, then such integration can save employees a lot of time and thus provide better service to customers. Where CTI is implemented server-side, there is less of an issue as it is relatively easy to add a network interface to the server that bridges the two VLANs/LANs. However, you may have network security engineers up in arms if you implement such a bridge!

Ensuring quality

Should you decide, for whatever reason, to carry voice and data traffic on the same network, then there is one network service you should implement if you are to avoid voice performance issues. We are assuming that you have already ensured that the network is at least 100 Mbps end-to-end, and preferably Gigabit. However, this alone is not enough, particularly if you are working to an SLA. Fortunately, this issue has been recognized and addressed through the **Quality of Service** (**QoS**) protocols.

Quality of Service is simply, as its name suggests, a means of ensuring quality on an IP network. A typical IP network will carry traffic for many different services, whether they be very much in the background (NTP or Network Time Protocol might be a good example of this) or quite obvious to the end user (HTTP for instance).

While most of these services can tolerate some delay as packets are re-sent, occasionally a network may run a service that is time-critical, such as voice traffic or streaming video. Using QoS on managed switches or routers, time critical traffic can be given a relatively high priority so that, if bandwidth is limited, such traffic is delivered before lower priority packets. There may also be mechanisms for allocating a portion of the available bandwidth solely to certain services.

If you are required to offer your customer an SLA on a network that shares voice and data traffic, then you should insist on QoS being implemented. Otherwise, the first time a user downloads a large file from the Internet, voice quality will plummet. It is also worth remembering that, to be truly effective, QoS needs to be implemented throughout the extent of the route that is bearing shared traffic. In other words, there is little point in having QoS on your LAN if, once the traffic is passed on to your ISP, there is no QoS for the rest of the route. The easiest means of addressing this problem is to have a dedicated voice circuit outside the LAN. For a small business this may mean one ADSL line for voice traffic and another for other Internet traffic, for instance.

There are two approaches to QoS—IntServ and DiffServ. IntServ allows a very fine level of control of QoS parameters, where you can determine the quality levels for each individual flow of traffic. IntServ requires that all routers that could potentially carry traffic between the nodes on a network are compliant and store all the configuration information for each traffic flow on the network. There is a significant overhead involved with this approach as the number of nodes grows, which obviously does not scale well to a network the size of the Internet, and so is not a common approach for VoIP installations.

The more common QoS implementation—DiffServ—does not offer quite the same level of control, but will scale much better. It requires all traffic to be categorized into a number of classes, which then have the quality rules applied to them. Usually, the classes are rated from 0 (lowest quality) to 7 (highest quality). In a router, packets are examined to determine which class they are in, and held in one of a number of queues until their 'turn' has come to be transmitted. This can actually result in an increase in network traffic if too many low-priority packets are being held in queues until they expire, resulting in the need for re-transmission.

A successful end-to-end QoS implementation requires that all routers in the path are configured to treat each class of service in the same way. Therein lies the problem with most commercial VoIP implementations, as it is usual for the customer to have little or no say in what happens to traffic, regardless of which class of service it is tagged with, once it leaves the LAN and starts traversing the ISP's network and the Internet. It is also possible that an ISP's customers would try to gain an advantage by categorizing all their traffic as highest quality. Therefore, most ISPs will apply their own rules to traffic traversing their network, a technique known as **traffic shaping**.

As a maintenance task, it is wise to review all the classes each time you make a significant change as the addition of another stream of traffic into the VoIP class can have a hugely detrimental effect on call quality.

Bearing all this in mind, is it still worth implementing QoS within a LAN? I would have to say that if you are carrying voice and data traffic on the same LAN then it is, mainly because it will prioritize voice traffic between the PBX and the phones, and also ensure that outbound voice traffic is presented to the ISP before less time-sensitive traffic. In particular, it will ensure that a user downloading a large file will not suddenly grab all the available bandwidth. But you, and your customer, should be aware that it does not guarantee high quality external calls, it merely mitigates some of the risks. As previously stated, a dedicated voice circuit is the easiest means of addressing this problem (outside the LAN). For a small business, this may mean one ADSL line for voice traffic and another for other Internet traffic, for instance.

When things go wrong

However well a network is designed, and regardless of the quality of the components, one certainty is that somewhere along the line something will go wrong. Good design and components may push that point back in time somewhat, but it will happen. It might be a device on the network that breaks, it might be a cable that works loose or gets bent, or it might be a switch port (or even a whole switch) that fails. At this point, the time and effort you spend on developing plans to deal with such issues will determine how badly the business is affected by the network issue. If you have no plan then you will be launched into a frenzy of high priority activity every time an issue occurs, however small. If you have foolishly agreed to an SLA with no provision for dealing with such issues, then you are bound to end up breaching it.

All is not lost, however, as putting a provision plan together is basically about applying common sense. Firstly, you should consider where you might experience problems; the answer is everywhere, but break it down into logical areas such as phones, cables, switches, routers, Internet circuits, PBX, and so on. Then look at the impact to the business of a failure of a device in one of these areas, categorizing devices into logical groups if necessary (such as reception phones, helpdesk phones, admin phones). At this point you should have a manageable number of scenarios that you can consider for likelihood and impact to the business.

Don't make the mistake of focusing on potential issues, rather than looking at effects. For instance, have a plan for dealing with the failure of a desktop phone, and then worry about what caused the failure after the service has been restored.

For each scenario you should then be able to discuss contingency plans with the customer. Simple Red/Amber/Green (RAG) charts can be useful at this point to visually demonstrate the risks that can be accepted and those that need addressing, either to reduce the likelihood or to improve the speed of response thus limiting the impact to the business. In general, the response to different scenarios fits well with RAG categorization.

Red

Normally, it has a high business impact and high or moderate likelihood. Any such scenarios should be addressed before implementation by improving the resilience of the system through the use of failover devices or similar measures.

Amber

Typically, it has a moderate impact/likelihood, or low impact/high likelihood. Often the best plan for these situations is to ensure the ability exists to restore service quickly. This might involve keeping a store of replacement devices (phones, cables, even a switch or two) or having an appropriate SLA with a service provider (for example a 1 hour response/ 4 hour fix for Internet outage).

Green

Usually, it has a low impact and moderate to low likelihood. Rather than invest money up front in dealing with such scenarios, customers usually prefer to deal with them as they occur. For instance, an admin person's phone failure could be dealt with by diverting calls to their mobile while a replacement phone is ordered, or the failure of a switch port could be dealt with during working hours by moving the cable to a free port, allowing you to wait until a low-usage time to swap out the switch.

Increasing resilience

Should your risk assessment highlight any "Red" risk areas, the most likely action required to deal with that risk is to increase the resilience of your system. Usually that means introducing redundant equipment or components to the network with the purpose of "stepping in" in case of failure. Approaches that can be used to increase resilience of the network and the telephone system are discussed in Chapter 7.

Summary

The success or failure of a Voice-over-IP implementation frequently relies on the quality of the network carrying the voice traffic. That quality is usually measured in terms of bandwidth, latency, and jitter. When implementing a new VoIP system, it is important that the existing network infrastructure is understood, and that any changes that need to be made to ensure the success of the implementation are communicated and agreed early on, particularly if there is a requirement to provide an SLA to the customer. The network is required to carry voice traffic, frequently alongside other traffic, in a timely manner if call quality is to be acceptable. Implementing Quality of Service throughout the network may be crucial to effective prioritization of voice traffic. It may also be required, should the customer deem it important, to deal with loss of components without a break in service, known as **failover**. This is essentially achieved through the use of redundant equipment running alongside the active devices. Planning and agreeing all this work up front with the customer is essential to the ongoing success of the system.

3
Call Routing with Asterisk

In the last two chapters, we have looked at some advanced dialplan techniques, and the means of ensuring that your customer's IP network is VoIP-ready. Spending time and effort on this, however, is wasted if you do not consider what happens once the call traffic leaves the customer's network. Unless you are routing calls within the LAN or WAN, normally meaning internal calls or inter-office calls, the call will be required to leave your control and pass to a service provider. This needs careful consideration up front as which type of service you use, and when, will affect the cost and quality of each call.

Using an IP-capable telephone system opens up a raft of call routing possibilities that can easily appear quite bewildering, even to someone who has already implemented an Asterisk-based system. However, by breaking your requirements down into bite-sized chunks it is possible to navigate this potential minefield and ensure you are routing all your calls effectively. This chapter explains the options available for call routing, and suggests some techniques that can be used to improve the performance of your Asterisk system, both in terms of cost efficiency and call quality.

Routing methods

Until the 1980s, telephony service providers were difficult to tell apart. After all, telephony traffic was all carried the same way, using copper cables, so there was no real need for any niche operators. Therefore, the telephony world was populated with a number of monoliths, such as AT&T in the USA, BT in the UK, and many others around the globe. In many countries, these providers were nationalized and thus were pure monopolies. In countries where this was not the case, competition was not strong and markets tended to be carved up so that monopolies or cosy cartels were the order of the day. In the US, this was evidenced by the antitrust case "United States versus AT&T", which in 1982 led to the breakup of AT&T into a smaller core company and seven regional Bell companies.

Mobile telephony, introduced in the 1980s but not widespread until the following decade, shook things up slightly. However, the massive investment required to provide an effective service worked against independent start-ups gaining significant market share, and the net result was that the same monoliths simply diversified into another route to market.

IP telephony is now changing the market significantly. Suddenly it is possible to make a phone call without touching the traditional telecoms network, although it is true to say that the old telephone companies are still at the forefront of Internet circuit provision services. However, newcomers do not require quite the same level of investment to gain a foothold, so the customer is offered more choice and market forces drive prices down and the standard of services up. There is still some way to go, particularly as the majority of calls are required to break out of the Internet and on to the PSTN or mobile infrastructure at some point. It should be pointed out, though, that this is not necessarily a bad thing. The PSTN, in particular, is still a low-cost, high-availability network.

Out of the box, so to speak, Asterisk is designed to work with IP-telephony protocols such as SIP and IAX2. The only routing requirement once it is installed is to have a connection to an IP network, often the Internet but not necessarily so. It is perfectly feasible to have your Asterisk box hooked up to a private WAN with no external routes. Your call destination options would be limited to the extent of the network, but you'd have great functionality and call quality!

In the real world, though, people like to make phone calls outside their LAN/WAN. One option is to set up an account with an **Internet Telephony Service Provider** (**ITSP**), pass all outbound calls through that trunk, and let the ITSP worry about how to route your calls. This has advantages in that it is a very simple setup (in all systems simple is a good thing) and it is usually the cheapest means of making international or long-distance calls. The ITSP will route the call over the Internet to a point near the call destination, where it will break-out onto the PSTN or mobile infrastructure. In this chapter, we want to look at adding a bit more complexity, controlled using some of the techniques discussed in Chapter 1, so that we increase our routing options with a view to reducing call costs, increasing resilience and ensuring quality. This means adding the ability to route calls from our Asterisk system through to different service providers, typically PSTN and mobile networks.

Where to start

Before you start adding new routes to your customer's Asterisk system you should, of course, sit down and take stock of what your customer's call profile is like at the moment. How many calls are made in a typical month? How many of them are local, national, international? What are the destinations of the international calls (and the national calls too if this makes a difference to the call cost)? What hours of the day, and days of the week, experience the call volume peaks and troughs? How long does the typical local call, national call, and international call last for? In other words, you need to understand what is happening to the calls within your company, or your customer's company if you are a consultant.

You should also try to gain an understanding of the costs and options available to you from telephony suppliers. For instance, at the time of writing, a major telephony provider was offering the following call package to small businesses with an annual spend of over $500:

- Local and National calls charged at 7c per minute
- Local and National calls up to 60 minutes capped at 20c
- Calls to national mobiles/cell phones up to 60 minutes capped at 50c

Of course these are headline figures and don't show such items as the per-minute rate to mobiles/cells or any international call charges. There is also a line rental charge to factor into the equation. However, these figures do give a flavor of the type of deal that is available. When carrying out this exercise for real you should, of course, see what deals are available from a number of suppliers.

Taking the example above, we can see that a customer that makes lots of local or national calls with an average duration of 40 minutes might very well spend less than a customer of an ITSP who offers such calls on a flat rate of 1c per minute. In fact, until recently, it hasn't been unusual for the cost comparison between PSTN and VoIP providers for local or national calls to come down in favor of the PSTN provider, a fact recognized by VoIP providers who now offer similar bundles of calls at low prices in an attempt to grab a share of that particular market sector. Offers of unlimited national and international calls (to a restricted list of countries) for about $20 per month can currently be found.

When carrying out these exercises, there is always a temptation to get drawn into a "what if" discussion. "What if" the company doubles in size over the next three years. "What if" the company's customers prefer calls to mobiles more and more. In truth, there are any number of "what if" questions. However, it is difficult to make an informed decision when enough detail is not known. Bear in mind that altering the profile of your call routes in Asterisk is not an onerous task, so changing circumstances can be catered for as they occur.

Thus, by now you should have a good idea of how many calls the company makes, the main destinations and the times of the day, or night, at which they are made. With this information you can decide what external routes you need to consider. Usually they will be a combination of the following:

- Internal same site
- Internal other site
- Local land line
- National land line
- International land line
- National mobile/cell
- International mobile/cell

Internal calls

These calls are by far the easiest to route. For internal calls in the same office it is simply a case of using the LAN and we have already discussed how to ensure your LAN is up to the task in Chapter 2.

If you are making inter-office calls, then the best route depends on the quality of the company WAN, if there is one. A WAN with good bandwidth between offices can have a single central Asterisk server handling all the extensions. It's a simple setup, and all resilience efforts can be focussed on a core server. However, losing connectivity to an office will mean that all telephone services at that office are lost. This risk can be alleviated by introducing resilience into the WAN links too, a measure which increases ongoing circuit costs.

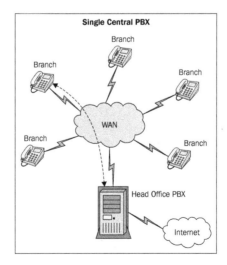

As we can see from this diagram, traffic from the branch office traverses the WAN to reach the PBX at the head office where it will be routed via the Internet if appropriate (for example if the call is external and to be routed via an ITSP).

Alternatively, a multi-server setup can be implemented where each office has a local Asterisk server, which can route all outbound calls via the Internet and/or PSTN if connectivity to the central Asterisk server is lost.

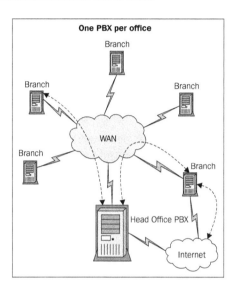

In this diagram we can see that each branch office now has a PBX. Calls can now be made internally in each office whether or not the WAN link is available. Call traffic can be routed from PBX to PBX, or as shown with the bottom-right branch, directly from their PBX to the Internet if needed or desired. For internal calls, this can be used as a failover route if the WAN circuit is lost. For the sake of clarity the link to the Internet for each branch has been omitted, but it can be assumed that each office could, and probably would, have an Internet circuit in addition to a WAN circuit. Of course, the WAN link can utilize the Internet circuit, with WAN traffic traversing a VPN tunnel.

So far we have assumed that the customer will have a robust WAN in place. A company with a poor or non-existent WAN will be best served either by routing call traffic through the Internet to a central Asterisk server, or by installing an Asterisk server at each site. A consideration as to which solution suits the company's needs best is the route that incoming calls take. If there is a central DDI for all offices then a central Asterisk server is an obvious solution. If each office has its own number(s), then there may be an argument for having servers on site. Either way, to secure traffic, a site-to-site VPN service should be considered.

Companies with a high number of internal calls across offices may also need to increase the bandwidth at each office, or install circuits dedicated solely to voice traffic. As discussed in the last chapter, this is also a good way of ensuring other Internet traffic does not steal bandwidth from the voice traffic.

The best solution really depends on the customer's needs and budget, and quite often is a mixture of both, whereby large branches may have their own PBX and small ones utilize a PBX that resides in another office.

Local calls

It's not unusual for VoIP providers to struggle to provide value for money over a traditional PSTN service for local calls. In many markets there has been keen competition in this area that has resulted in a reduction of charges to the point where many very competitive deals can be found. For customers implementing an Asterisk-based system making many calls to local landlines, the answer frequently is to break out on to the local PSTN immediately rather than route via an ITSP. Of course, the calculation should be made in each case, as at lower volumes the cost of line rental can become significant. For instance, a company that tends to make many local calls during a short period of time (either a small part of every day or during a short period of each week or month) may end up needing a high number of landlines for a relatively small overall number of calls.

For customers where using PSTN for local calls makes sense, there are a couple of options. Probably simplest is to fit one or more PCI cards with either analog or digital interfaces into your Asterisk server(s). Conversion from IP to analog or digital is carried out by the server, increasing the use of its processing resources but removing such traffic from the LAN. Alternatively, PSTN gateway devices exist that sit on the network and interface with analog and/or digital PSTN lines, presenting the traffic to the Asterisk server, usually in the form of SIP trunks. They carry the advantage that they will not fail should the server do so. In this fashion, they can form an important part of resilient solutions.

Integrating Asterisk with analog and digital PSTN lines is discussed in detail in Chapter 9.

National calls

The solution for national calls to landlines follows the same decision process as for local calls to landlines, although of course the numbers will be quite different. Whilst PSTN providers will tend to charge more for these calls, ITSPs are more likely to have a flat rate for local and national calls, or indeed bundle them into an attractive package. Therefore, you may find that routing via your ITSP becomes a more attractive option in this case. This should have the knock-on effect of reducing the number of PSTN lines required as a whole, saving online rental charges.

An area not yet considered is that of calls to mobile telephones (cell phones). ITSPs and PSTN providers have tended to exclude these calls from bundles, although as we can see from the earlier example, this situation is changing. However, for the best deals on calls to mobiles you frequently have to look at the mobile providers themselves. It is typical for a mobile provider to offer a large amount of included call-time for a relatively modest monthly fee, bringing the cost per minute down to 2 or 3 times the cost of a landline call.

With cost differentials like that, it frequently makes sense for customers to consider routing calls to mobiles directly onto the mobile network. For smaller companies, the means of achieving this is through the use of GSM gateways. Companies with high volumes of calls to mobiles have other options, such as a direct circuit to the mobile provider. There are many different types of GSM gateway, some designed to integrate with traditional analog PBXs, and others that are aimed solely at the VoIP age, presenting the GSM channels as SIP trunks. They also vary in size from those that will accept a single SIM card to rack-mounted devices that will take dozens or even hundreds.

For an even smaller implementation, it is possible to avoid the cost of a GSM gateway by use of the `chan_mobile` channel driver (see: `http://www.chan-mobile.org/` for more information), which is a third party add-on to the standard Asterisk package. The channel driver allows the use of mobile phones as FXO devices, connecting to the Asterisk server through the use of Bluetooth. This driver also allows a Bluetooth headset to act as an FXS device. It is difficult to recommend this solution for use in a commercial environment, though, reliant as it is on the proximity of a mobile phone to the server and Bluetooth as the communication medium. This is especially so given that the one-off cost of a dedicated gateway device can be under $100.

International calls

Traditionally, international calls are where VoIP has provided significant and obvious cost savings over PSTN and mobile telephony. The continued popularity of Skype in a domestic environment, and in many commercial environments too, as a low-cost option for international phone calls is testament to the acceptance of VoIP as a mature solution in this area. Indeed, it can be said that Skype has done more to promote the use of VoIP across the board than any other product. As an Asterisk enthusiast or professional, it will no doubt have changed many of your initial conversations with potential VoIP users from an argument about the virtue of VoIP in general to a comparison of Asterisk with Skype. As a result, you are almost certainly au fait with the reasons for choosing a standards-based system over a proprietary one, and for giving options over choosing the most appropriate route in the background against making that choice overtly.

Back, though, to routing international calls — the choice when configuring your Asterisk server tends to be between different ITSPs, depending on which country you are calling. For many customers there is quite a limited list of countries to which they place significant volumes of calls. Sometimes you can find an ITSP that gives competitive rates for all these countries, in which case your international routing is relatively simple. In other cases you may find that using two or more ITSPs are necessary to gain the maximum call savings. From a routing perspective, this is obviously more complex. However, it can make it easier to increase resiliency, allowing you to failover to another ITSP if your preferred option is not available for calls to a certain country.

Occasionally, if many calls are made to a less than popular destination for VoIP traffic, it can be more cost-effective to use the PSTN in combination with calling card providers. This usually requires that an account be set up with the provider, then a prefix be placed before the dial string to inform the PSTN provider to route the call appropriately. All this can, of course, be carried out relatively easily on your Asterisk server, although in a commercial environment it is strongly recommended that some measure of confidence in the long-term viability of the service provider be sought.

Routing international calls to mobiles/cellphones is, at the time of writing, almost invariably cheaper through an ITSP due to the relatively high charges placed on international calls by mobile providers. In Europe, these charges have come to the attention of the EU and the providers have been obliged to reduce the charges somewhat. However, compared to landlines and VoIP, mobile telephony remains a much higher cost option outside of the call time bundled into contracts, and international calls are invariably excluded from those attractive bundles. Hence, it is often cheaper to route these calls over the Internet or the company WAN, and then break out into the local mobile network. If the company has a local office then it can be effective to install a local GSM gateway, depending on such considerations as network bundles, cross-network charges, and so on.

Alternative options

Of course, as more and more companies and individuals utilize VoIP telephony, an alternative to breaking out onto PSTN evolves. To illustrate, consider this hypothetical example. Company A, in the UK, has invested in an Asterisk PBX in order to reduce call charges and introduce functionality beyond their previous telephony system. They have a supplier in the US, Corporation B, who also has an Asterisk system for pretty much the same reason. They both use ITSP's to route their international traffic, although they do not use the same ITSP.

Most often, calls between A and B would be routed as follows:

A → ITSP → PSTN → ITSP → B

But the break out to PSTN somewhere near B only results in the call being routed to B's ITSP before hitting B's Asterisk server. Far more sensible, and effectively free, would be to route as follows:

A → Internet → B

No call charges, and a minimum of conversion—one means of achieving this would be to trunk the Asterisk servers together, and for companies that have a close working relationship, this is very viable. A, knowing the extension number of a contact at B, can simply dial a prefix that uniquely identifies the call as being for a contact at B, followed by the extension number, and the call is routed over the trunk without touching an ITSP or the PSTN. For a more detailed look at how to set this scenario up, have a look at the following article on the `voip-info.org` web site:

`http://www.voip-info.org/wiki/view/Asterisk+-+dual+servers`

The problem, of course, is in scaling this solution up. As company A grows, it will have close relationships with more and more suppliers and customers, and so will want to route calls to them without incurring call costs too. From an Asterisk setup point of view, this is manageable for a while, but there is also a need for the companies involved to share internal extension lists and so on. Additionally, this works best if both companies are using an Asterisk PBX, which is a laudable ambition but unfortunately does not reflect the reality. However, any PBX that allows you to set up a SIP trunk (or even better an IAX2 trunk) should be able to connect over the Internet to an Asterisk PBX.

Luckily there are a couple of other options for routing inter-company calls via the Internet alone that do not require them both to be Asterisk systems, or even for either to explicitly define the other in their configuration.

ENUM

ENUM is a means of translating a phone number into an IP address that your Asterisk server will recognize and can use to route calls directly over the Internet to another VoIP-capable PBX. It works using the same technology that translates a URL typed into your web browser into an IP address that identifies the location on the Internet of the web server. That service is, of course, the **Domain Name System (DNS)**. For ENUM to work, a company needs to define its public phone numbers as it would a URL, linking them to the public IP address of its PBX. You can, and should, also define the protocols accepted by the PBX (for example SIP, IAX2).

DNS registration can be carried out internally or with an external registration authority such as `e164.org` or `e164.arpa` (ENUM DNS zones are usually referred to using the e164 prefix as e164 defines how and by whom telephone numbers are assigned). The roll-out of ENUM across the globe is progressing, albeit quite slowly, so there may be a local ENUM registrar for your region or not. If your intention is to use ENUM as a central store of outbound routes to external PBXs, and you are not concerned about allowing others to connect to your PBX through the Internet, then internal DNS makes sense. However, enlightened suppliers and customers will appreciate a public listing.

Asterisk allows you to define multiple ENUM lookups, so you can check internal DNS zone(s), followed by one or more external ENUM DNS zones. Although there is an argument for doing a local ENUM lookup last, just in case you define a catch-all local entry which will cause the ENUM lookup to fail before reaching the public zones.

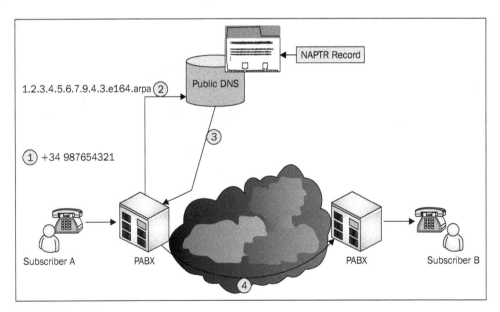

The preceding diagram illustrates how ENUM works by giving an example: Subscriber A sets out to call Subscriber B.

1. The User Agent of an ENUM-enabled subscriber terminal device, or a PBX, or a Gateway, translates the request for the number +34 98 765 4321 in accordance with the rule described in RFC 3761 into the ENUM domain `1.2.3.4.5.6.7.8.9.4.3.e164.arpa`.

2. A request is sent to the Domain Name System (DNS) asking it to look up the ENUM domain `1.2.3.4.5.6.7.8.9.4.3.e164.arpa`.

3. The query returns a result in the form of so called **Naming Authority Pointer Resource (NAPTR)** records, as per RFC 3403. In the previous example, the response is an address that can be reached in the Internet using the VoIP protocol, SIP per RFC 3261.

4. The terminal application now sets up a communication link, and the call is routed via the Internet.

> Basic ENUM Lookup: See `http://en.wikipedia.org/wiki/Telephone_Number_Mapping`
>
> RFC 3761: See `http://tools.ietf.org/html/rfc3761`
>
> RFC 3403: See `http://tools.ietf.org/html/rfc3403`
>
> RFC 3261: See `http://tools.ietf.org/html/rfc3261`

To use ENUM for routing calls from your Asterisk server, ENUM simply needs to be defined as a trunk. As routing calls via ENUM is effectively free, you may want to set this up as the first route tried for all calls to landlines. When Asterisk attempts to route calls via ENUM, a DNS lookup of the phone number is made, and if that returns an IP address and protocol, then an attempt is made to place the call via that route. If the lookup fails then the call should failover to the next route.

```
exten => S,1,Set(enumresult=${EnumLookup(${EXTEN})})
exten => S,2,Dial(${enumresult}) ;; lookup was successful
exten => S,3,Congestion
exten => S,52,Goto(${enumresult}|1) ;; got a TEL record, so forward
exten => S,102,Congestion ;; lookup failed
```

Within the ENUM DNS records, it is possible to have multiple destinations, as is the case with MX records for emails. Therefore, if the first destination fails, then the next can be tried, and so on. This is quite an elegant means of implementing a simple call forwarding solution.

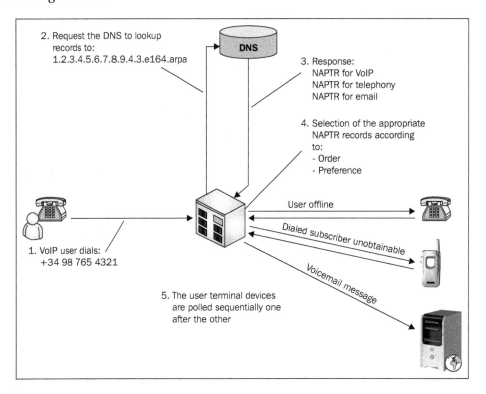

This figure graphically represents call forwarding using ENUM.

DUNDi

DUNDi (Distributed Universal Number Directory), devised by Mark Spencer of Digium, avoids the use of a centralized DNS registry for ENUM records by employing peer relationships in order to share dialplan information. In order to get over the lack of a single (or at least a very small number) point of reference for addressing information, DUNDi allows a PBX to query peers to see if they have the address information of the server hosting the required telephone number. Should the peers not have the information, they can query their peers, and so on until either the information is retrieved or the process fails. That failure can be due to running out of peers or by hitting the TTL, which is the number of iterations of the whole process (for example, a TTL of 2 for a particular query allows the peers of peers to be queried, but no further).

Using DUNDi in preference to ENUM has been compared to asking friends and colleagues for information instead of looking up a central information repository such as a directory. The first time you ask a question, you may get a quicker response from a directory (ENUM), as your friends/colleagues (DUNDi) have to ask their peers if they don't know the answer themselves, and those peers may have to ask others. However, the advantage of this approach is that once a question has been asked and the answer found, it is then "remembered" so that subsequent requests for the same information can be answered directly.

DUNDi-capable servers cannot do very much in isolation, so an explicit peering relationship with at least one other DUNDi server is needed. Public/private keys are used in the peering process, generated using the ASTGENKEY command. Each DUNDi server can choose which context(s) it advertises, allowing it to share all or a subset of its numbers with other DUNDi peers. A detailed description of how to set up a DUNDi peering arrangement between two boxes can be found at the following URL:

`http://www.voip-info.org/wiki/view/Asterisk+DUNDi+Call+Routing`

Once you have your peering arrangement in place, you can test it using the following CLI commands:

DUNDi show peers

DUNDiLookup number@context bypass

The first command will show the status of the peer(s) you have set up. The second will attempt to locate the specified number in the specified context (bypass ignores the cache, allowing you to determine if the number is currently available). If a route is found, you will get a response similar to this:

```
1.       0 IAX2/priv:ByWFbOGKgGmZbM43BJHSZw@192.168.1.2/301 (EXISTS)
         from 00:0c:29:d2:d8:ec, expires in 3600 s
DUNDi lookup completed in 113 ms
```

The response consists of 6 parts:

1. `IAX2`: The communication protocol.
2. `priv`: The context.
3. `ByW[...]HSZw`: The secret key of the PBX, used to redirect you to `ext 301`. This key changes on a regular basis to aid security.
4. `192.168.1.2`: The PBX's IP address. Normally this will either be a domain name or a public IP address.

5. `301`: The extension, which may differ from what you requested as the other party may translate an external extension number to a different internal one.

6. `EXISTS`: This number is being advertised by the PBX. This should not be taken as a guarantee that the extension is available as the PBX could advertise more extensions than are actually reachable.

More information on DUNDi, including white papers and best practices, can be found at the DUNDi home page:

`http://www.dundi.com/`

It is worth remembering that DUNDi and ENUM can easily co-exist side-by-side, as they are different mechanisms for achieving the same goal, which is to place a call to another organization across the Internet. For most external PSTN or VoIP calls you will probably want to interrogate DUNDi and ENUM for possible routes before using trunks that will incur a cost to the customer. This brings us nicely to the subject of how to implement such a routing strategy.

Types of routing

Once you have worked out the call profile, done the calculations to determine the best route for each type of call, purchased relevant gateways, installed and configured them, then you then need to set up the routing within Asterisk. With routing beyond the simple "take all outbound calls and send them through this trunk" it can be easy to tie yourself in knots. Where you have identified different routes, and therefore different trunks, for the types of calls already mentioned, perhaps with a different route being used depending on the time of day, then it's all too easy to end up with a hugely complex pile of virtual spaghetti.

It is recommended that, before any work is done on the Asterisk server, a lot of planning is carried out in close conjunction with the people who will end up using the server. As always, every implementation is unique, but certain rules of thumb should help keep the routing relatively clear and concise while still achieving the goal of minimizing call costs. Bear in mind, though, that it is more important for the system to place a call first time, every time, than to save a fraction of a penny, cent, zloty, and so on. If in doubt, keep it simple to begin with, prove that it all works, and increase complexity in manageable steps.

The general principles that should apply regardless of complexity are:

- Never have a single route for any one type of call. If a particular trunk is not available then you should always be able to failover to another one.

- Mix communication channels within your failover routes. Assume that any one type of route can fail, in particular, you should always assume that it is possible to lose Internet access and thus, connection to your ITSP(s). Therefore, failover to PSTN or mobile should be built-in.

- The 80/20 rule applies to cost/complexity ratio too. You will get 80% of your cost savings for 20% complexity and effort. Really think about whether the extra 20% of cost savings is worth the extra complexity, with its increased maintenance effort and likelihood of failure.

- Standardize as much as possible. If you have a lot of potential outbound routes, then it is worth considering having one or two primary routes for each type of call, and using one or two standardized failover routes. For instance, you might decide that whichever route is preferred for a particular type of call, if that is not available then you will execute a standard subroutine that will, in turn, try to route the call via an ITSP with reasonable charges to national and international destinations, then PSTN. This gives you one place to go to maintain your failover routes.

Routing techniques

Within Asterisk, as we saw in Chapter 1, call routing is achieved by matching the destination number against a mask. This relatively simple process can, with a bit of thought, achieve quite complex routing results. In general, the best principle to follow when setting up routing masks is to start with the obvious, simple rules, and gradually work through to the more complex rules. For example, you may decide that you want to set up the following routes:

1. All emergency numbers (such as 911, 999, 112, and so on) via PSTN.
2. All local calls (dial code starting 0207 or 0208, or numbers without dial code) via PSTN.
3. All national calls via ITSP A.
4. International calls to USA/Canada (dial code starting 001) via ITSP B.
5. International calls to other countries via ITSP C.
6. National mobile calls (dial code starting 07) via GSM gateway.

Logically approached, the routing could be achieved thus:

1. If number is 911, 999, or 112 then route via PSTN.
2. If number starts 07, route via GSM gateway.
3. If number starts 001, route via ITSP B.
4. If number starts 00, route via ITSP C.

5. If number starts 0207, route via PSTN.

6. If number starts 0208, route via PSTN.

7. If number starts 0, route via ITSP A.

8. Otherwise, route via PSTN.

As local numbers without a dial code are the most difficult to match against a mask, it is easiest to match everything else first and leave the local numbers without a dial code to be rounded up in the "catch-all" at the end.

Hence, in order to implement the rules described above, using macros for code clarity, we should end up with something along the following lines in the extensions.conf file:

```
;
; Send emergency numbers to PSTN
exten => _999,1,Macro(out-PSTN,${EXTEN})
exten => _911,1,Macro(out-PSTN,${EXTEN})
exten => _112,1,Macro(out-PSTN,${EXTEN})
;
; Send numbers starting 07 to mobile
exten => _07.,1,Macro(out-GSM,${EXTEN})
;
; Send numbers starting 001 to Provider B
exten => _001.,1,Macro(out-prov-b,${EXTEN})
;
; Send numbers starting 00 to Provider C
exten => _00.,1,Macro(out-prov-c,${EXTEN})
;
; Send numbers starting 0207 or 0208 to PSTN
exten => _0207.,1,Macro(out-PSTN,${EXTEN})
exten => _0208.,1,Macro(out-PSTN,${EXTEN})
;
; Send numbers starting 0 to Provider A
exten => _0.,1,Macro(out-prov-a,${EXTEN})
;
; Send anything else to PSTN
exten => _X.,1,Macro(out-PSTN,${EXTEN})
;
```

Note that in the final "catch all" line, we don't simply use the pattern "_." as it matches everything, even the Asterisk special extensions like s, t, i, h, and so on. It is better to use the pattern _X. that will match on any two (or more) character dial-string, so long as the first character is a number between 0 and 9.

You can, of course, implement more intelligent pattern matching for this code if so required. The use of standard macros to handle each outbound route is recommended, though, so that you do not have to duplicate code for each rule.

The process for failing over to another channel is to use the DIALSTATUS variable, as per the example below, which shows the possible contents of the out-PSTN macro mentioned above. A return value of CHANUNAVAIL indicates that the channel cannot process the call and so you should failover to another channel.

```
exten => s,1,Dial(${ZAP/1/${ARG1},,T)
exten => s,n,NoOp( Dial Status: ${DIALSTATUS})
exten => s,n,Goto(s-${DIALSTATUS},1)

exten => s-NOANSWER,1,Hangup
exten => s-CONGESTION,1,Congestion
exten => s-CANCEL,1,Hangup
exten => s-BUSY,1,Busy
exten => s-CHANUNAVAIL,1,SetCallerId(${CALLERIDNUM})
exten => s-CHANUNAVAIL,2,Dial(SIP/sippeer/${LOCALAREACODE}${ARG1},,T)
```

In this code, the call is attempted via the PSTN initially, but if that channel is unavailable then the call is routed through a SIP channel instead. As we can see it is a relatively simple process, but as already intimated, proper use can result in a very resilient and frugal call routing strategy.

You could, of course, implement the failover strategy as a standard macro in its own right if your customer is happy to have all failovers handled in the same way. For example, regardless of which channel is tried first, failover to ITSP Provider A, followed by ITSP Provider B, followed by PSTN, followed by GSM gateway. It is easier to maintain than a separate failover section for each outbound route, but could potentially cost the customer a little extra over time. However, having a separate failover strategy for each outbound route is not very onerous to maintain and usually makes sense for the customer.

Summary

Devising a routing strategy for your customer first requires that you understand in some depth their current call patterns (local, national, international, mobile/cell, and so on), and the options available in the local market for PSTN, mobile/cell, and VoIP traffic. Only then can you devise a plan for an Asterisk-based system that will ensure quality and value-for-money. It may also be that a customer with multiple sites needs to implement or upgrade the WAN as part of the implementation, as well as consider carefully whether they should have a single central PBX, or multiple smaller ones.

Once all this information is known, strategies should be devised for local calls, national calls, international calls, and calls to mobiles/cell-phones. A primary route for each should be chosen, with failover to alternative routes in place to ensure that calls can always be made. This can be implemented in a clear and effective manner through the use of macros in the dialplan. Use of ENUM and/or DUNDi should also be considered as a means of routing external calls via the Internet to avoid charges from telephony providers (including VoIP providers). Both ENUM and DUNDi are means of advertising Internet-enabled telephone numbers so that other subscribers can set up a direct link between PBXs.

4
Call Centers—Queues and Recording

Asterisk as a product is amazingly rich in features. If you can think of the telephony problem, nine times out of ten you can provide a solution with Asterisk, but there a few areas where Asterisk is less than perfect. To be fair, no product can excel in all areas. All of them have their weak spots, and Asterisk is no different. The good news is that there are often acceptable workarounds.

This chapter explores the call center environment. Believe it or not, using Asterisk as a base, you can deploy a system using open source software for a fraction of the cost of some of the commercial offerings out there with equivalent capabilities. One Asterisk-based system that does exactly this is VICIDIAL, and we shall have a look at it over the coming pages.

First, though, we'll examine some of the issues particular to call centers with "vanilla" Asterisk, and then we'll explore the solutions.

Asterisk queues

Asterisk has rudimentary call queuing capabilities built into the system which work well to a certain extent, in the fact that they are functional, albeit with limited capabilities. If you simply want to ring a whole bunch of phones then that is what happens. There are also the following ring strategies (from the `queues.conf` file):

- ringall: ring all available channels until one answers (default)
- roundrobin: take turns ringing each available interface
- leastrecent: ring interface which was least recently called by this queue
- fewestcalls: ring the one with fewest completed calls from this queue

- random: ring random interface
- rrmemory: round robin with memory, remember where we left off last ring pass

Queue gotchas

The problem is that apart from the first option, the system will only ring one phone or agent when it encounters the others. If that phone is busy or goes unanswered, you would have to exit the queue and do something else. You can of course use cascading queues, but they can get messy very quickly.

Another gotcha is that if you don't set the call limit equal to one for each device, you're going to get call waiting tones played to the agents potentially on active calls, which is not good. And then of course you might have issues with call transfers as you only have one channel to work with, this is also not good. On Snom phones, you change the call waiting indication to visual only, which still allows call waiting functionality, but doesn't beep in the agent's ear.

The final point to consider is how queue timeouts work. One might expect that if you pass a timeout value when you call the queue function, it would exit when the timeout expired. This appears not to be the case. The passed timeout value is only checked once the timeout value within the queue's definition has expired. Confused? Let's explore the issue a little deeper.

Once a call is in a queue, it only cares about timeouts set in `queues.conf`. Therefore, you need to set a timeout value in each queue definition. So, if we set a value of `timeout=60`, the queue application will only resurface to check the passed timeout every 60 seconds.

The bottom line is—if you passed 30 seconds as the timeout when you called the queue, it won't exit until one minute has expired.

A practical queue

Reading the above, you may well feel that queues are pretty constrained in their application, however, consider this scenario—a busy executive wants any call hitting his desk phone to ring his desk phone, and at the same time call his mobile/cell. You might think it's easy ... we'll simply ring both the numbers:

```
Dial(SIP/200&IAX2/${mytrunk}/07749000001)
```

There is no problem if the mobile is switched on and has a signal, but if it's switched off, guess what? It's going to go to voicemail straight away and perhaps ring the desk phone once or twice.

Using queues to cascade calls

In order to resolve the above situation, what we need to do is ring the desk phone and delay the call to the mobile so as to give the executive a chance to answer the phone on his desk (if he's there).

You could of course do this:

```
Exten => s,1,Dial(SIP/200,10)
Exten => s,2,Dial(SIP/200&IAX2/${mytrunk}/07749000001)
```

The problem with the above is that you'll get a break in the ringing, so that if the desk phone is picked up (just as it's going to the second priority), it will result in a dead call.

A neater solution is to use the queue application to kickoff multiple calls for you with delays where required. Look at the following example of queues in `extensions.conf`:

```
; Call Sales
exten => 1,1,NoOp(calling sales)
exten => 1,2,Queue(salesQ)

[Queues.conf]
[salesQ]
joinempty = yes
member => Local/*35@call_nik_mobile
member => Local/1@call_sales/n
member => Local/1@call_200/n

[..extensions.conf]

[call_sales]
exten => 1,1,Dial(${SALES},30,ortT)
exten => 1,2,VoiceMail(200@default,su)

[call_200]
exten => 1,1,wait(5)
exten => 1,2,Dial(SIP/200&SIP/201,30,ortT)

[call_nik_mobile]
exten => *35,1,wait(10)
exten => *35,2,Dial(SIP/${mytrunk}/07749600000)
exten => *35,3,Hangup()
```

In this example, an inbound call is routed to `[salesQ]`.

Within [Queues.conf], we've defined three static members to call. The queue application executes all three of them at the same time. In essence, now we have three independent threads running for a single inbound call.

The net result is that the phones defined in the ${Sales} ring group are called, and ten seconds later, the extension 200 starts to ring. In the meantime, the sales phones continue to ring uninterrupted. Finally, the mobile starts to ring.

The above shows how you can stagger calls to devices without interrupting the ring process.

Call recording—the issues

These days when call recording is becoming a mandatory requirement for many business verticals, Asterisk falls at the second fence. There have been some improvements in the major version releases, but for anything over ten agents, there are major problems.

Although there is a patch available in Asterisk 1.6, recording data is currently written in 44-byte chunks. This is inefficient and has the following implications:

- File fragmentation: Writing such small amounts of data causes fragmentation, PC load issues, and hard drive stress. In .wav format, you're looking at 1MB per minute, so it's easy to imagine how many disk I/Os are going on.

  Files written concurrently will be extremely fragmented.

- Drive failure: Regular IDE and SATA drives are highly likely to fail with regular call recording use due to the sheer number of head movements.

So what are the solutions? Well for a start, if you record to a RAM drive, you'll relieve stress on the drive, and no longer suffer from fragmentation issues. You could also use SCSI or SATA-ES drives (which offer fast rates of throughput).

Show-stoppers

Even if you resolve the above, there is still a major problem. Using the generic call recording functions in Asterisk, recording will stop if you transfer a call! Why? That's because Asterisk uses local channels for call transfers. These are optimized to be destroyed when the call is transferred. When these temporary channels go away, so does the information that monitor/MixMonitor relies on, thereby the application terminates.

At the time of writing, there appears no way to surmount this issue, **unless** you use the native Asterisk method of transferring calls. This is done by hitting the transfer key (normally #) as defined in `features.conf`, which is pretty awful as you lose the nice transfer functionality of most SIP phones. More to the point, how do you ensure that the operator always uses native transfers? This is pretty impossible.

Ultimately, the most efficient way to ensure reliable recording is to relieve Asterisk of this responsibility completely. In other words, offload the recording functionality. Unless you have a low number of recordings going on concurrently (less than 10-15), the load on your Asterisk machine will start affecting the call quality.

VoIP recording approaches

With a packet-switched network, in other words Ethernet, the route that the voice traffic will pass through can be very difficult to predict, especially where routers are involved. This can cause a headache in developing a suitable solution.

Two basic approaches can be taken for recording calls in an IP environment:

- **passive**: Here the recording solution "sniffs" for appropriate data to record. You may have used Wireshark (Ethereal) to analyze SIP traffic. This product passively listens for traffic on a node of the network.
- **active**: This could be simply having a recording device in the handset wiring. In other words, you are actively forcing traffic via a recorder.

Impact of VoIP on recording systems

Let's take a moment to look at VoIP and recording systems.

Hardware convergence

Traditionally, voice and data have had separate sets of wiring. However, more and more commercial products now utilize CAT5 Ethernet connectivity. It's still common to have two separate networks where two providers are involved, but it makes sense to just have one. This is mainly because of the availability of switches with VLAN capability, which segregates the bandwidth and ensures that an appropriate level of bandwidth is reserved for voice traffic.

There are significant cost savings to be made by using the same connectivity standards. Although people may baulk at the thought of wiring telephone extensions using traditional telephony connections, running a length of Ethernet cable for a new phone can normally be handled with existing in-house expertise, thereby eliminating expensive phone engineer call-out charges.

Distributed call centers

A significant advantage of building a distributed call center is that it's easy to have personnel located in diverse physical locations, and yet be a part of the same logical call center. VoIP releases the shackles of being tied to a monolithic system.

Apart from the obvious reduction in costs of not having to house people in a large building with all the attendant travel, it's a fairly simple matter to make use of lightly utilized staff as overflow call center staff, such as those that might deal with back office matters. In these cases, recording nodes can be located in the remote offices, and managed centrally if they are not a part of the same physical LAN.

Home working

The ideal distributed call center is the one where individual Customer Service Representatives work from home with nothing more than a good broadband connection. This will allow them to view the client's data on screen as well as take the phone call, thereby removing the requirement of separate phone and data lines.

Recording could be achieved by using the local PC, but it would be even better at the central site that allocates the calls. VICIDIAL is a good example of a system that simply treats remote workers as long-wire extensions, and therefore captures the complete conversation.

VoIP recording challenges

As with all new technologies, there are some technical challenges to overcome:

Routing

In order to passively capture the voice traffic, the data must be reachable. It's no good that the recording system is listening on subnet 192.168.3.x, when all the calls are traversing 192.168.10.x!

Therefore, you need to ensure that your recording solution can see the traffic across all of the subnets you want to monitor. Sounds like an obvious statement, but you need to be careful – it actually happens if you have routers and switches connecting your subnets.

Bandwidth

Bandwidth is a valuable resource, with 100 megabit as the norm these days for internal cabling. So, as one might expect, internal calls are run using G.711. In other words, at full PSTN quality, ensuring excellent quality from a recording perspective.

However, what do you do about remote offices/workers that might be using compressed codecs? You're going to have full quality on one side, but degraded on the other.

Encryption

You can bet encryption is going to become a bigger and bigger issue, not only from a network security issue, but also for "finger printing" recordings to ensure they can't be tampered with. This opens up a whole can of worms that is yet to be resolved.

Solutions

While there are a number of commercial applications out there, Asterisk is all about open source and to that end, we'll look at an open source recording solution as well.

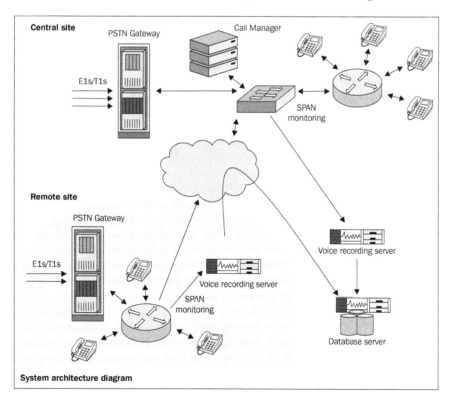

System architecture diagram

In the above figure, you'll notice that SPAN monitoring is a key element. What is it?

A few years ago, most office devices were connected using a hub. Any traffic sent to a port on the hub was broadcast to all of the connected devices. This would cause congestion issues on medium to high-level usage, as you would often get collisions. This means that the traffic would need to be resent.

The solution was the "switch". When a switch is powered on, and starts receiving traffic, it begins to build a Layer 2 forwarding table. It stores the MAC address of the sending device and the port that it is on, such that when traffic is received for that device, it knows which port to pass the traffic to. This means that once the table is built, traffic only passes between two ports and is not broadcast to all ports.

This means that traffic is sent a lot more efficiently, to the extent that hubs are pretty much obsolete these days. The downside is that if our recording solution is plugged into another port, it's not going to see the traffic! Enter SPAN.

In essence, a SPAN port is where all traffic is copied or reflected to a given port. Think of the `ChanSpy` application in Asterisk. The conversation is going on between two parties, but the spy application is SPANNING the traffic.

SPAN technology is now available in more and more switches, not just the high-end Cisco devices. If you're looking to do passive call recording, you need to look at your current switching devices.

OrecX (`www.orecx.com`) has been developing call recording solutions for some time. Their solution is of a passive nature, because it literally sniffs out conversations on the network and records them. The beauty of this approach is that there is no load on the Asterisk server. The open source element provides the backend functionality, but it's up to you to provide a frontend. Of course, OrecX can sell you a fully-featured frontend, and that's how they generate revenue to continue development. You will need to determine if the development costs of putting together a frontend outweigh the costs of buying the commercial solution. The choice is yours. That said, if you search for open source call recording, there's not a lot out there.

Call recording solutions sell for big numbers, and to date, it's been a fairly niche market. However, with more regulation on the way, requiring that voice recording becomes mandatory, you can expect to see more solutions being developed. But don't forget that you'll need to deploy SPAN-capable switches.

Asterisk call center solutions

There are a number of solutions out there, but the most complete one we've seen is VICIDIAL. This has been in development for a number of years now and continues to thrive.

In an interview with us, Matt Florell, the author, was asked what drove him to write this mammoth solution.

> *I started first working on astGUIclient during the summer of 2003. It was only an end-user interface at that point to facilitate click-to-record, call transfers, call logging, conference call handling and display of voicemail.*
>
> *It was originally built so that I could get used to working with Asterisk and the Asterisk Manager API as well as for use at a couple of client sites. Then the astGUIclient project was soon started on Sourceforge and the code was released as open source (GPL).*
>
> *Later in the year a one-lead-at-a-time dialer component (called VICIDIAL) was added for call center usage. This was done because a client had need for a dialer and was very surprised by the extremely high price of dialer systems on the market and only needed limited outbound functionality.*
>
> *In 2004, VICIDIAL became the focus of the project and basic multi-line auto-dial functionality was added. In 2005 the user interface changed from Perl/TK client-server to PHP/AJAX web-based to allow for easier use on agent workstations.*
>
> *In 2006 the predictive algorithm was added to the mix and from then on VICIDIAL development has added more and more advanced features to the program. Currently, there are over 1,000 installations of VICIDIAL that we know of in over 70 countries around the world and the agent user-interface has been translated into 10 languages.*

This product works very well, though it's not for the faint hearted. But let's not put you off just yet.

How VICIDIAL works

As you might expect, the system is based around a campaign. Within the campaign, you can set various parameters such as where this is an outbound, inbound, or blended (that is both outbound and inbound) campaign, how you want to recycle leads, and so on. The campaign is fed by leads which are read from the database and loaded into the virtual "hopper". A script processes this list and instructs Asterisk to dial the number. Once the call is answered, it is passed to an available agent along with additional data regarding the call, such as name, address, and so on.

VICIDIAL makes extensive use of Asterisk's conferencing module `MeetMe`. When an agent signs on, they are joined into a conference. Similarly, when the dialer gets through to a live customer, they are simply joined to the conference with the agent. As `MeetMe` sessions can be recorded, this gets around the whole issue of losing the recording on a call transfer as there isn't one.

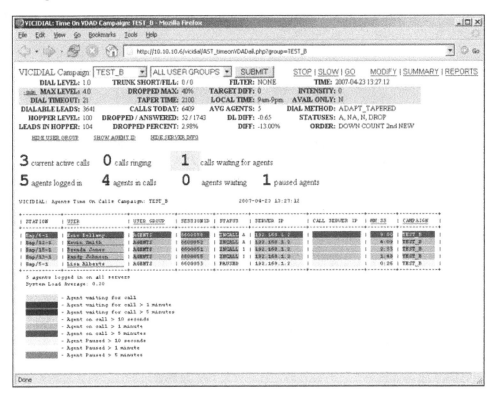

The screenshot above shows part of the administrator interface. As you can see, there are a number of agents in calls with one paused. As time goes by, the color changes, so you can instantly see who's been on a call for a long time. The screen also shows that there are three calls actively being dialed and one caller is waiting for a free agent. That's not good, because after a predetermined timeout, the call will be considered **DROPPED**. The regulatory authorities take a dim view of any operation that has a high-dropped call percentage, as they are considered nuisance calls. I'm sure you've experienced silent calls in the past, and these are caused by this type of a situation. Fortunately, VICIDIAL is FTC-compliant and has the appropriate mechanisms to cater for this eventuality. In this example, the dropped percentage is 2.98%. Once 3% is reached, it goes red to alert the admin that the dial rate is probably too high.

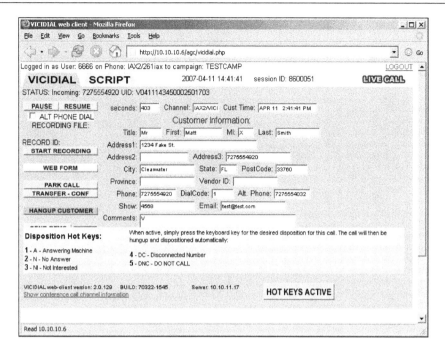

The screen above shows a typical agent's screen on a live call. When a call is connected to an agent, the form is populated with all of the information of the customer. Of special note in this example are the **Disposition Hot Keys**. They allow the agent to disposition a call with a single key press.

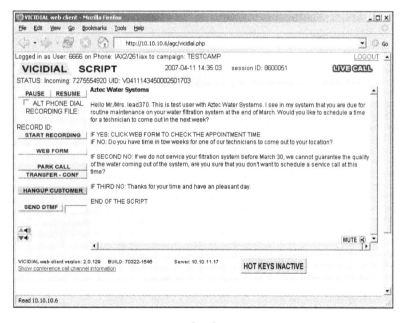

Apart from the standard screen shown previously, VICIDIAL has the ability to either launch a web form or a script on call connect. In the above example, a script has been launched. This can use variable substitution (also known as mail merge) to populate the script with real time information.

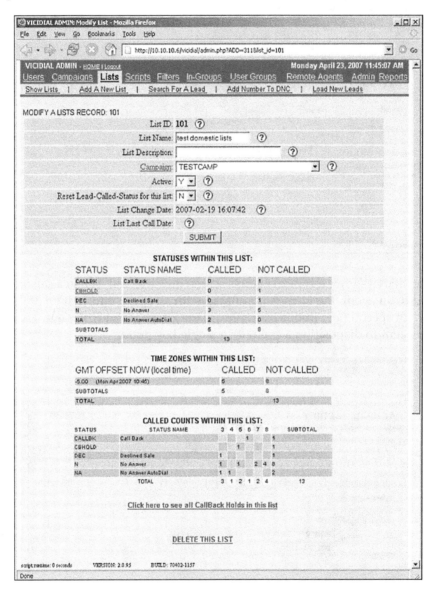

Lists are the heart of the system. They are allocated to a given campaign. In the previous screenshot, we're looking at the settings for a given list. We're shown the totals for each disposition, how many have yet to be called, and so on. Within a campaign, you can set lead-recycling parameters, such that if a given number is busy, then it can be redialed after a given preset period.

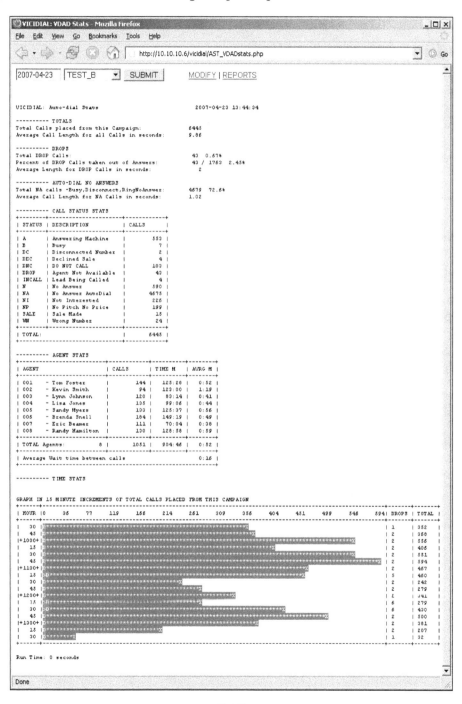

As with any system, there are a number of reports available. The previous one shows the statistics on a particular campaign. Along with the call statistics, you can see which agent had the most calls, and so on.

Handling inbound calls

So far, we've mainly looked at outbound call handling. One of the bug bears of running a call center is return call backs. In the UK, those called have the tendency, out of curiosity, to dial 1471 to call back a missed call. If you're regulation compliant, you should be presenting caller ID on your outbound calls to allow this. The problem is that your call center could be overwhelmed by these callers. You can take two views on this, namely:

- We don't want to be bothered by these calls
- We see this as an opportunity

Either way, VICIDIAL has a solution for this. In the first instance, you can route the inbound call to a message stating the reason for the call. In the second, you can inject the call to the live agents.

The next screenshot shows a real snapshot of a working call center. As it's just started up, the dropped percentage is high, but remember that the figure to aim for (3%) is over the day. What this screen shows is that we have two agents on outbound calls (denoted by the **A**) and one handling an inbound call (**I**), one paused, and two waiting for calls.

The first call denotes an inbound call yet to be answered. Within a split second, this call will be passed to one of the two waiting agents.

This example shows call handling for inbound/outbound using the same set of agents, but you could equally have separate campaigns for in/out with different sets of agents. The choice is yours.

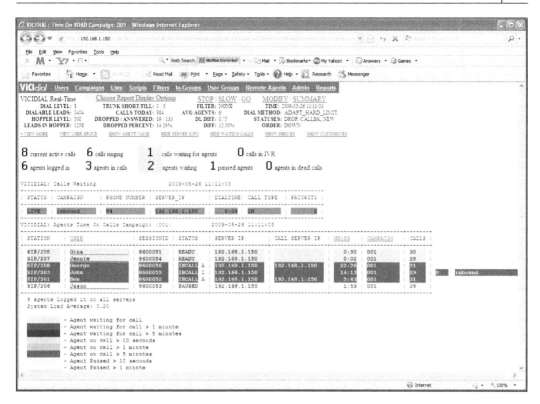

Installation

As mentioned at the beginning of the section, installing VICIDIAL is not for the faint hearted. One of the key recommendations is to build a custom kernel. One of the primary reasons for this is to optimize Linux as a server. In other words, we don't want to use it as a desktop machine, so processor preemption needs to be turned off, Most distributions out of the box have the kernel built to handle both desktop and server mode, which is a compromise. There's a whole bunch of other optimizations listed as well, timing being absolutely essential, as we're running the Asterisk `MeetMe` application which relies on a good timing source. The VICIDIAL scratch install is a useful guide to the installation, but it needs to be read carefully. Don't try to install the application on a remote machine unless you really know what you're doing. Make your first installation on the machine you can lay your hands on quickly. If you're not careful, you're likely to lock yourself out with a dead kernel. That said there's an excellent, if lengthy, "Scratch install" document online, which will take you through the steps one line at a time.

Timing sources

The Asterisk MeetMe application is highly dependent on a reliable timing source. Until recently, the only viable option was to provision a Digium TDM card. Setting aside the cost of the card, a "real" issue was the availability of a suitable slot to plug it into. Bearing in mind that you need a **server** as opposed to a **PC** to run VICIDIAL under any sensible load, you can find yourself restricted to using ztdummy which is far from ideal. For whatever reason, when a Digium card cannot be used, there are now devices specifically manufactured to supply robust timing through a USB port, which are pretty ubiquitous on servers these days.

Scalability

VICIDIAL can be separated into three basic components — the database, the web server, and the dialer. Smaller setups, those having less than 20 seats, can all be run on one server. Start moving beyond that and you need to start separating those components. As a rule of thumb, it's a ratio of 1:2:4 — one database server, two web servers and four dialers. With each dialer capable of handling 25 agents, you now have capacity for 100 agents. Working on an expenditure of $1,500 per server, the total hardware solution comes in at less than $10,000. That may seem a lot, but try getting a quote anywhere near that from any of the larger commercial vendors.

Summary

This chapter has discussed an Asterisk-based call center solution (VICIDIAL) and shown how you can build a scalable and cost-effective solution for your customers. We also discussed some of the shortcomings of recording with Asterisk and solutions to work around them. In addition, we also looked at the inbuilt call queuing application, its pitfalls and limitations, as well as demonstrating how to use queries to provide delayed/cascading ring functionality.

In conclusion, while "vanilla" Asterisk can struggle to deal with call center environments, in a highly-tuned system such as VICIDIAL, coupled with an external call recording solution, Asterisk can form the cornerstone of a compelling solution.

5
Asterisk and Speech Technology

Telephone systems that can speak and recognize what the caller says may seem like technology gone mad, but there are a number of compelling reasons why such implementations may make sense. After flirting with speech technology for a number of years, the telephony industry is now at a point where the "name dialer" is a fairly common application. Until recently, things like **automatic speech recognition** (or **ASR**) were the preserve of those vendors with very large R&D budgets and those users with similarly deep pockets to buy the systems.

Once again, Asterisk has served to democratize this technology, along with many other telephony applications, by drastically lowering the barriers to entry.

First we will consider why speech-enabling telephony applications which currently work with DTMF might be a good idea. Following that we will identify and briefly discuss the three main types of speech technology in abstract, before looking at what kinds of integration are possible with Asterisk, and walking through some detailed installation and implementation examples.

In this chapter you will learn the following:

- Good reasons for speech-enabling your telephony applications
- The three types of speech technology that are used in the telephony world
- What to consider when speech-enabling your solution
- How to painlessly add speech recognition to Asterisk
- How to easily add text to speech to Asterisk
- Implementation advice and tips

Why speech-enable?

Telephony applications such as the auto attendant (or IVR) have been around for a long time and obtaining user input through the DTMF keypad has proved successful enough—so why go to the trouble of enabling users to say what they want rather than "Press 1 for ..."?

Speech is one of our natural interfaces. We get a lot of training in using speech from a very early age and, it is argued, telephony systems with a speech interface are more customer-centric because of this. People, while conscious that they are interacting with "a system", feel more at home having that interaction in one of their natural modes as opposed to being conformed to the way the machine wants them to do it.

Although "warm and fuzzy" feelings for the user are important, there are also a number of very practical reasons why deploying a speech interface with your solution could be a wise proposition, depending on your target market in terms of types of customer and geography.

Allowing a user to speak to a system removes the need to use a keypad, so it is much safer for applications that may be used when people are mobile. When someone rings to check their voicemail while driving, it is best to allow the process to be as "hands-free" as possible. Systems like this have been in use for some time in a number of safety-critical environments. For example, some up-market cars allow the driver to issue basic commands by voice like "lights on". The aviation industry is also a user of speech technology, here cockpit warnings and alerts are "spoken" to the pilot by a speech synthesis system.

There are still countries where DTMF is not widely deployed. Some South American countries have a large population of loop-disconnect dialing phones—and while it is possible to detect loop-disconnect dialing, it takes a lot more resources to do that than it does to detect DTMF, and that can make scaling an issue. In such geographies it may be best to implement a speech-enabled system from the outset.

Speech recognition is also of great benefit to users who, for one reason or another (visual impairment, unable to use fingers, and so on), are not able to use the telephone keypad.

Perhaps the most compelling and practical reason to speech-enable certain types of telephony application is the fact that you can dispense with long, multi-level menus—"Press 1 for ..., 2 for ..., 3 for ...", then "Now press 1 for ..., 2 for ..." Hooray!

Types of speech technologies

There are three main types of speech technologies. These are **Automatic Speech Recognition (ASR)**, **Text-to-Speech (TTS)**, and **Speaker Verification and Identification (SVI)**.

Automatic Speech Recognition (ASR)

ASR is the process of a system listening to the user speak, and then (hopefully) coming up with the words that were spoken. There will also be a confidence score returned from the recognition engine that indicates the degree to which the system is sure of its guess.

Let's have a look at various flavors of ASR that are available.

Isolated Word Recognition

This is where users can only speak a single word at a time, such as "sales" or "support" or "play" or "stop" — this type of recognition is sometimes known as DTMF replacement, as it does not really add anything that did not exist before — it just changes the interface by which commands are given.

Connected Word Recognition

This allows the user to speak a collection of words (usually two or three) to be recognized. The recognizer will then analyze the utterance in order to establish what it thinks those words were. A process known as **directed dialogue** is used to help the recognizer by telling the user (through the prompt) what sort of thing they are expected to say. An example would be `<prompt>` `"Say the first name and the last name of the person you wish to speak to"`. The user then knows the type of thing to say. The recognizer is programmed to be looking for two words, say, and will compare what it "hears" with a list of first names and last names, such as `David Duffett` or `Jared Smith`. The collection of words that a recognizer is trying to "listen out" for is called the **grammar**. Naturally — the smaller the grammar, the greater the chance of an accurate recognition.

Natural Language Recognition

This describes a scenario where a user is greeted with a prompt that says something like, "How can I help you today?" Obviously, there is no real way of knowing how a user might respond and so the job of the recognizer is to "understand" every word uttered and, by the use of a sophisticated algorithm, come up with an idea of what the request is. Natural language recognition systems have been successfully deployed but, as you would imagine, they tend to take a huge amount of time to refine and can still be somewhat fragile.

You may have come across speech recognition software for use on desktop computers. Let's consider, for a moment, the differences between the desktop "dictation" scenario and the one we are really interested in, on the end of a telephone connection, as this will help us to understand some of the challenges that we will experience during implementation.

Desktop Speech Recognition (dictation)	Telephony Speech Recognition (IVR)
Requires a period of training where the user reads out known sentences in order to optimize the recognition process.	Must make a "guess" the first time it is fed an utterance — users would not expect to have to train the recognizer.
A separate profile is created (including the "training" process) for each speaker.	Must be speaker-independent.
Usual input would be a reasonably good multi-media PC mic.	Usual input would be the microphone within a phone.
Between the microphone and the recognition software is a short cable and a sound card.	Between the microphone and the recognition software is the phone circuitry, a telephone network and the interface from that into the host machine.
The user can instantly see what the recognizer thinks has been said (on the screen of the PC).	The user has no way of knowing what the recognizer thinks has been said, unless this feature has been designed as a part of the application. The desired result (for example, the call being directed to "sales") is achieved, or not.

Language coverage in ASR is absolutely essential as it is a language-dependent technology. It must be ensured that the language of the users is covered by the ASR provider chosen for a project. The only exception is for local variations of a language. For instance, if your market is Australia, but your ASR provider only has US English or UK English, you will probably be able to tune one of those to do a good job for Australian English.

Text-to-Speech (TTS)

The ability for a system to "speak" given text has been around in one form or another for many decades. You may remember being able to have your early home computer "say" things, or the Texas Instruments children's toy "Speak & Spell". As with ASR, the exponential rise in processing power available and the developments in TTS methodologies have meant that it can now be part of telephony systems.

While a recording will always have the edge in sounding natural, TTS is more than acceptable for giving information over the phone, which is subject to rapid change—making recording impractical. Actually, most telephony TTS is the playback of lots of small recordings. These sounds include **diphones** (the transition between one sound and another, like the a-v in David), and by concatenating lots of these little chunks of sound, full words can be made. Depending on the specific application, whole words and phrases will be recorded too, enabling the things that will be said most often to sound natural.

For example, reading out stock or weather reports will involve heavy use of certain words such as (in the case of the weather report) "rain", "sunny", "wind", "coming in from the", "north", "east", "south", "west". In these cases the range of things to be synthesized from small chunks of sound is reduced because whole words and phrases can be used for a lot of the time. This makes the process faster, and the output sound more natural. In telephony applications things can be a lot more general—for example, reading out emails. In the case of reading out emails, some of the content of an email is not straight text. The header of an email will need special treatment, for which the TTS system will use its text pre-processor.

TTS is also, of course, language dependent and so careful attention must be paid to the language coverage of the TTS provider chosen for a solution.

Speaker Verification and Identification (SVI)

Speaker Verification and Identification is the speech technology that allows the system to tell whether you are who you say you are! Do not fall into the trap of confusing this with *speech* recognition (where the system is interested in what is said), this is all about *speaker* recognition (who is saying the words, not the words themselves).

Speaker verification is the process of validating an identity claim—someone claims they are David Duffett, so the system will compare their voiceprint with that of David Duffett to see if there is a match. Please note that the term **voiceprint** can be somewhat misleading—in the vast majority of systems this is not a recording but a mathematical analysis of the voice constructed during the users' enrollment to the system.

Of course, speaker identification is a more difficult proposition because it is no longer a case of testing whether someone is who they say they are—it is about establishing identity. This means going through all the voiceprints on the system, possibly thousands, and coming up with the best match(es).

For this reason, applications today are much more about speaker verification. A prime example of this type of application is the password reset package. Did you know that in a company with thousands or tens of thousands of staff, the IT helpdesk spends many months every year resetting passwords (to a known default) for users who have forgotten them? By integrating a speaker verification application with their back office software, enrolled users can ring up, voice-verify, and then get back into their system to choose a new password. The payback time on such systems is claimed to be in terms of months.

Before you rush out to sell your customers an SVI password reset system, it should be pointed out that, at the time of writing, there are no known SVI packages that work with Asterisk, but it has the potential to be quite a cool money-making application. Maybe you are the person to write an SVI package, or integrate an existing package, with Asterisk? SVI is language-independent, so the same SVI package will work in any geography.

MRCP

The **Media Resource Control Protocol** (or **MRCP**) is not a speech technology in its own right, but a standard interface through which speech technologies can be controlled. Most ASR, TTS, or SVI companies will have (or be able to point to) an MRCP "wrapper" for their products. Asterisk does not currently have an MRCP capability and so only ASR and TTS engines that have been specifically integrated with Asterisk will work at the moment. MRCP is just mentioned here, as anyone looking into speech technology will come across the term.

Implementation considerations

So now you know we are going to home in on ASR and TTS for use with Asterisk. Both of these can be quite hungry in terms of processing power and memory usage—be sure to use some reasonable horse-power if you are serious about implementing speech technology. At least 2GB RAM with a 2GHz CPU would be a good start.

For this reason, most providers of speech engines use a client/server architecture to allow speech processing to occur on other (perhaps more powerful) machines than that on which the telephony application is running—although this will not always prove necessary.

ASR and Asterisk

There are a number of options for adding speech recognition to Asterisk and these include Sphinx and LumenVox. For our example, we will go with LumenVox as it makes use of the native speech recognition API (application programming interface) that has been part of Asterisk since version 1.4 was released.

This API allows providers of speech engines to hook their software into Asterisk in a way that means it can be controlled by a standard set of speech-related dialplan applications. Thus, those developing Asterisk speech applications need only to learn one set of dialplan applications, regardless of which (compliant) speech engine is doing the work behind the scene. We will go through these dialplan applications later.

LumenVox do provide a lot of support for Asterisk users and we would like to thank them for making a lot of resources available through their website www.lumenvox.com.

Installing LumenVox speech recognition with Asterisk

There are few pre-requisites for installing LumenVox speech recognition with Asterisk:

1. A PC with the processing power and RAM to handle doing the speech recognition in addition to doing the other work it has to do (in this case running Asterisk, as we will be adding LumenVox to an existing Asterisk machine).

2. A Linux distribution supported by LumenVox. Currently Fedora Core, Red Hat Enterprise Linux, CentOS (through Red Hat support) and rPath (Pound Key/Asterisk Now).

3. Asterisk 1.4.11 or newer (and all dependencies) installed and running properly—this example is based on Asterisk 1.6.0.3.

4. If using Red Hat ES or CentOS, LumenVox also requires 'libjs' (the Mozilla implementation of JavaScript) which, if not already on the machine, can be obtained as source from the official Mozilla website. A number of sites offer a compiled RPM of the library – and although not certified or supported by LumenVox there are more details on the LumenVox website. One such site that hosts the packages (at the time of writing) is http://dag.wieers.com/rpm/packages/js/.

5. An internet connection—the more bandwidth the better as you will be downloading about 150 MB!

Once these pre-requisites are in place we can begin. We will use CentOS 5.x as our example—other distributions will follow a similar pattern (Red Hat will be identical).

We need to obtain the LumenVox software (there are several components) and to get started with Asterisk a single channel "starter kit" is available from both the LumenVox and Digium web sites. At the time of writing it costs $50 which, to add sophisticated speech recognition to your Asterisk server, is a complete bargain! Implementing LumenVox on a production system will cost more, but is still excellent value compared to other offerings in the telephony space. You will need one license for each concurrent channel of speech recognition you wish to run.

The LumenVox software comprises of the recognizer itself (LumenVox SRE), the licensing server, the Asterisk connector (`res_speech_lumenvox.so`) and the Grammar Editor tuning utility.

So, with our starter kit purchased, let's go:

Log on to your machine as root—you must be root to install LumenVox. We are going to use `yum` to do our package management, so we need to get the `gpg` key, in order to build a repository from the appropriate packages.

```
# cd /etc/pki/rpm-gpg (this is the directory we want to download it into)
# wget http://www.lumenvox.com/packages/EL5/i386/RPM-GPG-KEY-
LumenVox.gpg
```

Now we are going to set up a LumenVox repository in the correct directory. This will tell `yum` where to go on the LumenVox web site for the packages.

```
# cd /etc/yum.repos.d
# vi LumenVox.repo
```

Here are the contents for this new file:

```
##################################################
[LumenVox]
name=LumenVox Products $basearch
baseurl=http://www.LumenVox.com/packages/EL5/i386/
enabled=1
gpgcheck=0
##################################################
```

Write the file and quit the editor.

We are now ready to install LumenVox. There are number of components that we will require. The available packages are:

- LumenVoxCore that contains core files shared across all packages. This is required.
- LumenVoxClient that contains the speech client.
- LumenVoxSRE that contains the speech server.
- LumenVoxLicenseServer that contains the License Server.

Typical installations will require everything to be installed. To download and install everything, you could run the following command:

```
# yum install LumenVoxCore LumenVoxClient LumenVoxSRE
                                    LumenVoxLicenseServer
```

The products we install will register themselves as services in /etc/init.d/ called lvsred (the speech server) and lvlicensed (the License Server). By default, they will be started automatically when you login. They are not started by default after installation—you must either start them using the service name start command or by logging out and back in after installation.

Now we need to log out and then log back in (as root) to update the environment variables, among others.

 It is possible to install LumenVox without using yum (although this is probably the easiest way). More information is on the LumenVox web site.

Our next task is to set up the licensing to enable the speech recognition engine to run. The license will be tied to the MAC address of the host PC, and LumenVox supplies a utility to generate a file which must be uploaded to the LumenVox web site. This in turn generates a file that we can then download to our machine—completing the licensing process.

First we need to be sure that the Licensing Server is running:

```
# ps -e | grep lv
```

It should show the following two services:

```
lv_sre_server
```

```
lv_licence_server
```

If you cannot see the licence server, try:

```
# service lvlicensed start
```

Next we will use the command: `/usr/bin/lv_license_manager` with the `-g` `<path/Info.bts>` which tells the licence server to store the hardware information for the machine on which it is running into the `<path/file>` specified. For example, if we wanted to create the server ID file in the root directory, we would type:

```
# /usr/bin/lv_license_manager -g /Info.bts
```

This will generate a file called `Info.bts` in the root directory. Log into your account on the LumenVox web site, and click on the **View Licenses** link in the **My Licenses** box. In the **Tools** part of the deployments section, you will see an **Upload License** link. Clicking on this will let you choose a name to identify the computer on which you are running LumenVox and it will prompt you to browse to the `Info.bts` you just created.

Once you have uploaded the `Info.bts` file, you will find that the link that used to say **Upload License** has magically changed to, guess what — **Download License** — because some smart software on the LumenVox web site has now generated a license file for your machine. When you click on **Download License** you will be prompted to review and agree to the license terms and conditions and, if you click agree, you can then download your license file to your chosen location. The file will be called something like `licensexxxx_xx.bts` where xxxx_xx will be some numbers with a letter near the end of the sequence.

Now type `/usr/bin/lv_license_manager -m Licensefile` where `Licensefile` is the path to the license file you just downloaded. The license manager will then install the license file, such as:

```
# usr/bin/lv_license_manager -m /root/Desktop/License2344_r2.bts
```

The `-m` option means merge.

You should see a message letting you know that the operation was successful. Hence, LumenVox is installed and the license server is enabled. We can check that both processes are running like this:

```
# ps el | grep license
# ps el | grep LV
```

Both of these should show that the appropriate process is running.

If you would like to check that the speech engine is running before going on to integrate it with Asterisk, LumenVox supply some sample code (called `example.cpp`), which needs to be compiled, and so on similar to the following:

```
# cd /usr/share/doc/lumenvox/client/examples
make

./example 127.0.0.1
```

This will run `example` which passes an audio file of someone speaking the number 8587070707 to the speech recognition engine. It outputs what has been recognized which, of course, should be 8587070707! The output will look similar to the following:

```
count=0, decode returns 10
Interpretation 1:
8587070707
```

If it does, then the speech engine is working and we can move on to integrating with Asterisk. This program loops, so you will need to *Ctrl+C* to get out of it.

To connect LumenVox to Asterisk, a **connector bridge** is used. This module hooks LumenVox up to Asterisk's native speech API and is downloaded as a tarball from the (**My Downloads** part of the) LumenVox web site. It could be downloaded to `/usr/src/` or anywhere you choose to use. Our first step is to "untar" it:

```
# tar -zxvf asterisk-1.6.x-lumenvox-support-linux-i686-32bit-b23-
engine8.6.tar.gz
```

This will expand the files zipped inside the tarball into a sub-directory of the current directory named after the file — so we need to `cd` into that new subdirectory:

```
# cd asterisk-1.6.x-lumenvox-support-linux-i686-32bit-b23-engine8.6
# ls
```

This will show four files, two of which we are particularly interested in — the connector bridge itself and `lumenvox.conf`.

```
# mv res_speech_lumenvox.so /usr/lib/asterisk/modules
# mv lumenvox.conf /etc/asterisk
```

This first `mv` moves the connector module into the directory where all Asterisk modules are stored. The second `cp` puts the `lumenvox.conf` file into the directory where all Asterisk configuration files reside.

To get Asterisk to load its generic speech module and the `res_speech_lumenvox.so` module we must edit `/etc/asterisk.modules.conf` — inside this file we will find (a good few lines down) the line:

```
;preload => res_speech.so
```

It must be uncommented (by removing the ";") and we must add a line to load the LumenVox module — it should look like this:

```
preload => res_speech.so
load => res_speech_lumenvox.so
```

Save the changes and quit the file. We can use the defaults set in `lumenvox.conf`, so there is no need to edit this file right now.

All we need to do now is to restart Asterisk (if it was already running) and we're ready to go. We can then go into the Asterisk CLI and check the presence of the relevant modules by:

```
CLI> module show like speech
```

We should see:

res_speech.so	Asterisk generic speech module
res_speech_lumenvox.so	LumenVox speech module
app_speech_utils.so	Asterisk speech-related dialplan applications

If all three are present, we're in business. To check that things are working from within the dialplan, LumenVox have some "quick to add" lines for the Asterisk dialplan on their web site—here is one set:

```
[lumenvox-test]
exten => s,1,Answer()
exten => s,n,Wait(1)
exten => s,n,SpeechCreate()
exten =>
  s,n,SpeechLoadGrammar(yesno,/etc/lumenvox/Lang/BuiltinGrammars/
  ABNFBoolean.gram)
exten => s,n,SpeechActivateGrammar(yesno)
exten => s,n,SpeechBackground(beep)
exten => s,n,Verbose(1,Result was ${SPEECH_TEXT(0)})
exten => s,n,Verbose(1,Confidence was ${SPEECH_SCORE(0)})
exten => s,n, SpeechDeactivateGrammar(yesno)
exten => s,n, SpeechDestroy()
```

All that is needed is to add these lines in and then add a line at an appropriate place in your dialplan to direct a call to the `lumenvox-test` context, for example:

```
exten => 3333,1,Goto(lumenvox-test,s,1)
```

The above example uses a very simple grammar file to look for the words `yes` or `no`. The recognized word is displayed on the CLI along with a confidence score (out of 1,000).

If the `safe_asterisk` script is being used, some environment variables must be added to the script—full details can be found at `www.lumenvox.com/asterisk`.

This set of steps will get us to the point that LumenVox is installed and working with Asterisk, but we still need to design our own dialplan to do what we want to do! To that end, we will now go through the example in step 12 (above) to find out more about each of the speech-related dialplan applications.

Checking that things are working

After we have answered the call and waited for one second:

SpeechCreate()

This application starts any speech session, and is mandatory as it creates the information that will be used by all following speech-related applications. It can take the speech engine name as a parameter, but if it is omitted it will use the default speech engine.

SpeechLoadGrammar (yesno,/etc/lumenvox/Lang/ BuiltinGrammars/ABNFBoolean.gram)

This application loads the grammar file for the channel (it is only operative for the life of the channel), which includes the actual words to be recognized, the order in which they might be spoken, and the result to return when a recognition occurs. The two parameters are a "label" and the path to the grammar file.

SpeechActivateGrammar(yesno)

This application activates the grammar identified by the "label", which is passed as a parameter (in this case yesno).

SpeechBackground(beep)

Similar to the dialplan application Background(), this application plays a sound file (the name of which is passed as a parameter) and tells the recognizer to start "listening" on the audio feed in.

Verbose(1,Result was ${SPEECH_TEXT(0)})

This line uses the Verbose() application to output the result of function SPEECH_TEXT with the argument 0 to the Asterisk CLI. The SPEECH_TEXT(n) function return contains the text (or semantic interpretation, if applicable) that a caller said. Here, n represents the number of results in case there are multiple returns from the Engine. The first result is number 0.

Verbose(1,Confidence was ${SPEECH_SCORE(0)})

This line uses the `Verbose()` application to output the result of function `SPEECH_SCORE` with the argument 0 to the Asterisk CLI. The `SPEECH_SCORE(n)` contains the confidence score for result n.

There are a number of other speech-related dialplan applications and a function, not used in the example above, which we need to be aware of.

SpeechStart()

This application tells the recogniser to start "listening" on the audio feed in, but unlike the `SpeechBackground()` application it will not play a prompt.

SpeechDeactivateGrammar(label)

This application deactivates the specified grammar.

SpeechUnloadGrammar(label)

After the recognition has taken place, one of the last priorities in the extension should be to call this application in order to avoid potential memory leak.

SpeechDestroy()

This application destroys the speech resources for the channel and therefore frees up the license.

${SPEECH(results)}

This function/argument will return the number of results.

Grammar files

In the small example above, the grammar file used — `ABNFBoolean.gram` — was supplied by LumenVox in the directory `/etc/lumenvox/Lang/BuiltinGrammars`.

For our own speech recognition applications, we may need to create our own grammars, so let's look at an example file:

```
#ABNF 1.0;
language en-US;
mode voice;
tag-format <semantics/1.0.2006>;
root $company_directory;

$mark = (Mark [Spencer]) {out = "200";};
$jared = (Jared [Smith]) {out = "201";};
```

```
$david = ((David | Dave) [Duffett]) {out = "202";};
$operator = (operator) {out = "0";};
$company_directory = ($mark | $jared | $david | $operator)
   {out = rules.latest()};
```

The above grammar file is written according to the Speech Recognition Grammar Specification, a W3C-defined standard for writing grammars. The file itself is in Augmented Backus—Naur Form (ABNF). It is an Internet technical specification format that is described in detail by the RFC 2234. The grammar is built on "rules".

All the lines in the file are necessary. The first four lines are a header, and you will notice that the kind of thing to be recognized is specified—it is "voice" and the language is "en-US"—be sure to use the most appropriate language for you target market. The next line (after the four-line header) is the root rule—it tells the recognizer which rule to begin with.

In this case the root rule is $company_directory, so the recognizer will start with this line:

```
$company_directory = ($mark | $jared | $david | $operator)
   { out = rules.latest() };
```

This rule is called company_directory and the rule expansion (which follows the first "=" sign) tells us that the recognizer is to "listen out" for $mark or $jared or $david or $operator. The { out = rules.latest() } tells the recognizer to return the "interpretation" instead of the actual word recognized. Hence, the $company_directory rule is, in turn, making use of four other rules.

Taking the rule $david as our example, the rule expansion means that the recognizer will understand that if the caller says David or Dave or David Duffett or Dave Duffett then the rule $david has been matched, and it will return the "interpretation", which occurs after the ":"—in this case 202.

Hence, when the caller says David or any of the other matches for the $david rule, they *really* mean 202 (David's extension number). The Dial() application can then be used to call the number returned by the recognizer.

Implementation advice for ASR

The first thing to mention is that the parser that goes through the grammar file for LumenVox is extremely "picky"—syntax, case, and space errors can result in all manner of different symptoms from a straight "grammar file not found" message (even though you have used the right grammar file name and directory, if the content is not up to scratch—you will be chastised!) all the way through to Asterisk restarting (if using the safe_asterisk script).

If you know LumenVox is running properly (from the previous tests described), the chances are that any errors you come across (other than "run-time" mis-recognition errors) will be connected to a problem with the grammar file.

The key to a great ASR implementation is to understand that making the system work is just the first 40-45% of the task. The ultimate success of the project will come by the tuning and optimization once the system has been deployed with your customer (your customer should be advised of this approach early on in the project). You should, of course, perform testing of your system in the lab before delivering it to your customer – but the tuning and optimization must be done in the customer scenario. To this end, LumenVox supply a tuning tool called the Grammar Editor (currently only available on Windows) so that you can tune and optimize the system based on real-life voice input from your customer's callers. It allows you to:

- import call data
- transcribe the audio data
- make adjustments
- test the changes and measure the results

It is an iterative process—tune, test, tune, test, and so on. You will need to agree upon a measure by which you and your customer can concur that the system does the job.

Keeping the number of words to be recognized (the grammar) to a minimum will increase the chance of an accurate recognition. The use of good prompts cannot be minimized when considering factors that can help ensure the success of an ASR implementation, for at least two reasons:

1. The friendliness, enthusiasm, intonation, and enunciation of the prompts will have a direct effect on the response of the caller and the clearer their response, the better the chance of a successful recognition.
2. The phrasing of the prompts is also a key factor, as directing the dialogue well will allow a smaller grammar to be used—again improving the chances of a successful recognition.

Remember the confidence score that the recognizer returns? We can use this to determine how well the recognizer thinks it has done and, if the score falls below a certain level (say 700 out of 1,000), we can use a confirmation to ensure the recognizer is correct or go back to the beginning of the recognition process. It is also possible to use 'weights' to help the recognizer know which words are more likely to be said by callers—more details on the LumenVox website.

One last thing–there is so much great information at www.lumenvox.com that it's a 'must visit' site. On top of that, the LumenVox people are very nice and helpful – and your starter kit includes free support to get it successfully installed (all for $50!).

TTS with Asterisk

As with ASR, there are a number of options for TTS with Asterisk. These include Festival, Cepstral, and Flite engines (all of which are available thanks to the initial academic research carried out at CMU). Our example will be based on Cepstral for a number of reasons, not the least of which is that with Cepstral, you can use the voice of Allison Smith—the voice of Asterisk. Since the job of a TTS engine is to "read out" the text it is presented with, integration with Asterisk is somewhat more straight-forward than integration with ASR.

Having said that, the actual task of reading out the text in a way that is understandable and acceptable is very complex, and maximum respect should go to Cepstral for what they have achieved.

The Cepstral web site has a great online demo facility which allows you to type text into a box and hear the various available voices speak it out. This way, you (or your customer) can choose the voice most fitting for a given application and geography. Remember to choose the 8 kHz version of the voice as these versions are specially adapted for telephony applications (other versions are dedicated to other applications such as desktop use). Assuming that Cepstral will be running on your Asterisk machine, the Linux flavor should be chosen.

With those choices made, a price will be displayed. At the time of writing, the price is less than $30 per channel! It is possible to download Cepstral TTS free for experimenting, but a license will be required for each concurrent channel in a deployment situation.

There are two ways of using Cepstral TTS—one method is to create the spoken text in a `wav` file that can be played in the Asterisk dialplan using `Playback()` or `Background()`. The other introduces a new dialplan application called `Swift()`, which speaks the text you pass as a parameter on the fly without intermediate steps. We will be using the second method.

Just in case you are wondering, **Swift** is the name of Cepstral's TTS engine.

Assuming that you have the Cepstral voice which you would like on your machine, let's go:

The file presented for download will be a tarball. Save it into an appropriate directory — /usr/src would be fine. Ensure that you're logged in as root, and then untar the tarball, now the contents will be expanded into a new sub-directory named after the file:

```
# tar -zxvf Cepstral_Allison_i386-linux_5.1.0.tar.gz
```

Change the directory into the newly created sub-directory and install the "voice":

```
# cd Cepstral_Allison_i386-linux_5.1.0
# ./install.sh
```

You will be prompted to agree to the license's terms and conditions and to the location of the installation, which is /opt/swift by default.

Continue further by typing the following three commands:

```
# echo echo /opt/swift/lib > /etc/ld.so.conf.d/cepstral.conf
# ldconfig
# swift "Hello World"
```

The first command creates a cepstral.conf file in the /etc/ls.so.conf.d/ directory from the contents of /opt/swift/lib.

The second command configures dynamic linker run time bindings.

After entering the last command, you should hear some real-life text to speech. If your Cepstral "voice" is not yet licensed, you will hear a meassge saying so, prior to the "Hello World" message.

The following command will show you the voice that you have installed:

```
# swift --voices
```

The resulting output will also show whether it is licensed.

Now we can add Asterisk dialplan support for Cepstral's TTS engine Swift by downloading an application called app_swift, created and maintained by Darren Sessions (to whom respect is due).

It can be obtained like this:

```
# cd /usr/src
# wget http://www.darrensessions.com/pub/app_swift/app_swift
1.6.2.tar.gz
```

Untar the tarball and the contents will be expanded into a new sub-directory named after the file:

```
# tar -zxvf app_swift-1.6.2.tar.gz
```

Change the directory into the newly created sub-directory, compile the code, and install it:

```
# cd app_swift-1.6.2
```

```
# make
```

```
# make install
```

At this stage, check for the presence of a `swift.conf` file in `/etc/asterisk/`. If it's there, great. If not, you can copy a file that was created when you "untarred" the tarball above:

```
# cp swift.conf.sample /etc/asterisk/swift.conf
```

Once you have a `swift.conf` file in `/etc/asterisk/`, you need to edit it in order to let Swift know which voice you have chosen — you will see the `voice=` line right at the end of the file, the default is `voice=David-8kHz`, change this to the voice you are using.

Your last job before getting back to Asterisk is to set up a link in your search path for Cepstral like this:

```
# ln -s /opt/swift/bin/swift /usr/bin/swift
```

At this point, you should restart Asterisk. You are now ready to test the presence of Swift in Asterisk and then edit your dialplan, adding a line to speak some text. On the Asterisk CLI, use this command:

```
core show application swift
```

If you get some information about the Swift application, all is well. Now you can add a line in your dialplan to speak some text, like this:

```
exten => 1000,1,Swift(your text here!)
```

You can register your Cepstral license by typing:

```
# swift --register
```

That's it — you're all done! You can now go ahead and implement many different TTS applications, a lot of which (these days) are based on pulling information from web pages or databases, and then reading it out. Some examples of such applications are:

- Weather reports
- Train or flight timings
- Tele-banking

Implementation advice for TTS

Here, you can do lots of testing prior to customer deployment. Try all the different things that the TTS engine might be expected to say to check for incorrect pronunciations. If these occur, it may be possible to misspell words in the text that you are using to make the pronunciation better. If the TTS engine will be reading out emails, try a few to see how headers are handled and so on.

Summary

Adding speech technology to Asterisk is relatively easy. However, you must be prepared for an extended development cycle that includes customer deployment in the case of ASR. Both ASR and TTS will demand reasonable processing power and memory, so specify these into the platform at the outset. Including speech technology could prove to be a key differentiator for you when it comes to your solutions, so become great at deploying it!

6
Call Accounting and Billing

A VoIP business built solely on installations will struggle to be viable, reliant as it is on the constant feed of new business to generate income. If you plan on growing your business, you need to provide (as far as is practical) an end-to-end solution. Then, when installations are hard to come by, you still have the regular income from call charges to sustain the business.

Billing is an essential element of that solution. This chapter discusses open source solutions to the billing dilemma as well as internal call accounting, allowing you to monitor your call activities as well as those of your customer's.

Call Data Records (CDRs)

Asterisk can store Call Data Records (CDRs) in a variety of formats. By default, call records are stored as .csv files. You could use an external program to process these files on a daily basis for analysis and they are great for a backup. Although, you'll probably store these in a database ultimately, but wouldn't it be great if Asterisk did it for you? Well you've guessed it—it can store them by way of the add-on modules.

Asterisk supports the following external databases for recording CDR data:

- SQLite
- PostgreSQL
- Any database that the Unix ODBC supports
- MySQL
- MS SQL or Sybase database through FreeTDS drivers
- Yada

We'll look at the MySQL implementation for this exercise.

Assuming that you've downloaded and unpacked the Asterisk add-ons package, the installation is similar to an Asterisk build-from source.

```
/configure
Make menuselect
Make; make install
```

Let's look at `menuselect`:

```
*****************************************
Asterisk-addons Module Selection
*****************************************


Press 'h' for help.

1.   Applications
                         --->
2.   Call Detail Recording
3.   Channel Drivers
4.   Format Interpreters
5.   Resource Modules


*****************************************
Asterisk-addons Module Selection
*****************************************


Press 'h' for help.

[*] 1.  cdr_addon_mysql

MySQL CDR Backend
Depends on: mysqlclient(E)
```

Note the `Depends` line at the bottom of the screen. If you have the following entry, you haven't installed the required dependencies:

```
[XXX] 1.  cdr_addon_mysql
```

Make sure you have MySQL installed (client and mysql-devel).

Before compilation, you will also need to edit the `cdr_addon_mysql.c` file and add the following to the top of the file so that a unique ID for the call is stored in the database:

```
#define MYSQL_LOGUNIQUEID
```

After you've installed the MySQL module, you'll need to create the appropriate table in MySQL.

At the command line, enter the MySQL client, that is, the username and the password:

```
mysql --user=root --password=mypassword
CREATE DATABASE asterisk;
GRANT INSERT
  ON asterisk.*
  TO asterisk@localhost
  IDENTIFIED BY 'mypassword';
USE asterisk;
CREATE TABLE `cdr` (
`calldate` datetime NOT NULL default '0000-00-00 00:00:00',
`clid` varchar(80) NOT NULL default '',
`src` varchar(80) NOT NULL default '',
`dst` varchar(80) NOT NULL default '',
`dcontext` varchar(80) NOT NULL default '',
`channel` varchar(80) NOT NULL default '',
`dstchannel` varchar(80) NOT NULL default '',
`lastapp` varchar(80) NOT NULL default '',
`lastdata` varchar(80) NOT NULL default '',
`duration` int(11) NOT NULL default '0',
`billsec` int(11) NOT NULL default '0',
`disposition` varchar(45) NOT NULL default '',
`amaflags` int(11) NOT NULL default '0',
`accountcode` varchar(20) NOT NULL default '',
`userfield` varchar(255) NOT NULL default '');

ALTER TABLE `cdr` ADD `uniqueid` VARCHAR(32) NOT NULL default '';
ALTER TABLE `cdr` ADD INDEX ( `calldate` );
ALTER TABLE `cdr` ADD INDEX ( `dst` );
ALTER TABLE `cdr` ADD INDEX ( `accountcode` );
QUIT;
```

Now that you've created the table, you'll need to edit the cdr_mysql.conf file. The following sample can be found in the config directory placed inside the addons directory:

```
; Note - if the database server is hosted on the same machine as the
; asterisk server, you can achieve a local Unix socket connection by
; setting hostname=localhost
;
```

```
; port and sock are both optional parameters.  If hostname is
specified
; and is not "localhost", then cdr_mysql will attempt to connect to
the
; port specified or use the default port.  If hostname is not
specified
; or if hostname is "localhost", then cdr_mysql will attempt to
connect
; to the socket file specified by sock or otherwise use the default
socket
; file.
;
; [global]
;hostname=database.host.name
;dbname=asteriskcdrdb
;table=cdr
;password=password
;user=asteriskcdruser
;port=3306
;sock=/tmp/mysql.sock
;userfield=1
```

So for our example, we configure the following:

```
[global]
hostname=localhost
dbname=asterisk
table=cdr
password=mypassword
user=asterisk
userfield=1
```

In order to load the module, restart Asterisk and make a test call. If all is well, you should now have an entry in your database.

CDR frontends

There are various web frontends out there, but the one that stands out is **Asterisk-Stat**. This is a simple-to-install, web frontend that supports MySQL as well as PostgreSQL. Full installation instructions are available at the following URL:

```
http://www.areski.net/areski/index.php?option=com_content&task=view&i
d=22&Itemid=54
```

Call accounting

You (or indeed your customers) may not necessarily require call billing on your Asterisk platform, but you may very well have a requirement for call accounting. Using a product like Asterisk-Stat provides many useful functions that not only highlight abuses of the system, but also provide useful information of how the system is utilized, such as:

- Who called who, when, and for how long?
- Who's making outbound calls and wasting time hearing the other end ring endlessly?
- If a customer claims they called you on a particular date, you can prove that they might be mistaken
- Which extension took a call from a customer?
- Which extension made a call to a customer?
- Who's making calls to inappropriate destinations or spending too long on a call?
- Are calls going unanswered because you don't have enough staff?
- Are you busier this month compared to last?
- To show the timings when you're busy

The graph above (Asterisk-Stat) shows the daily load on a system by the hour.

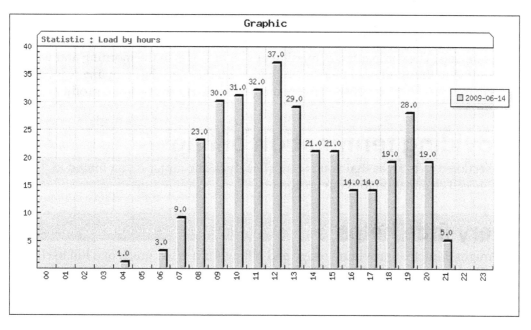

The next graph shows the call load on the system:

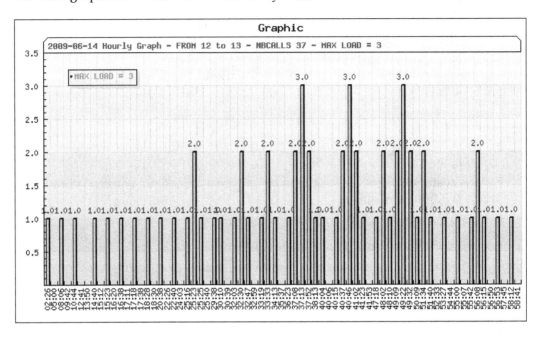

As we can see, the maximum load is three. In other words, in this 24-hour period, there was a maximum of three calls at any one period of time. How is this helpful? Let's look at a real life example:

For one client, we identified that on a Saturday morning, the maximum load was seven calls, but they only had five staff members. But in the afternoon, the load was four calls. The company was missing out on business in the morning and was overstaffed in the afternoon. By rearranging the staff roster, the company was able to increase sales, but their costs remained the same. The result—more profit!

Providing termination billing

This section covers areas that you are most likely to encounter if you intend to provide call termination for your customers.

Every little helps

You might think that providing usage and billing for it is too much of a hill to climb, but call revenue is what will make your business valuable.

Call revenue is measurable. As you add more customers, clearly your revenue will grow. And guess what, the more the calls that get placed, the better the "per-minute wholesale rate" you'll get. You may well feel that you'd rather pass on the hassle to another existing provider, and that's a fair point of view. However, if you do that, you'll lose control of the last leg of the call. If you point your customer at another SIP or IAX provider, and that provider has problems, who's the customer going to call first? You of course! You may or may not be able to help, but ultimately, you become a third party in resolving the issue and you're not even being paid for it.

Selecting a billing platform

As any web search for "Asterisk and billing" will prove, there are a number of solutions out there. So how do you choose? Good question. AstBill is the great looking package that it seems to be, but when choosing a solution, support is a major issue. Logon to the forum for AstBill, and you'll see pleas for help that go unanswered for weeks with posts that are months old. The open source community has many well-meaning projects that the original authors put their heart and soul into for months, sometimes years, before their will to carry on waned. Don't get me wrong, this is a fantastic effort. However, the development/support is no longer there.

Introducing A2Billing

A2Billing has been around for a number of years now, it's actively supported and a quick look at its forum (`http://forum.asterisk2billing.org`) shows posts everyday. At the time of writing, the current version of A2Billing is 1.3. The 1.4 version is in Beta and due for release in July 2009. The development team is open to new ideas and requests, and A2Billing 1.4 goes a long way to making this a carrier-grade product.

Reasons to consider A2Billing

- Support: It's well supported with lots of long-time contributors.

- Easy to install: The installation document is extremely well written and complete.

- Easy integration: Unlike a number of other solutions, it doesn't mess too much with your existing dialplan. You simply need to add a few lines of code and that's it.

- Scalable: You can point lots of Asterisk servers to a single instance of the database.

- Reporting: Clear concise reports showing call costs and profits are available.
- Invoicing: It enables sending invoices to your customers via email.

And the list goes on, but as mentioned previously, it's well supported!

A2Billing requirements

- Asterisk 1.2.24+ or Asterisk 1.4.0+
- Apache 2
- PostgreSQL 8.0 or MySQL 5.0
- PHP 5
- Suhosin security patch for PHP is strongly recommended
- PHP-PGSQL or PHP-MySQL
- php-pcntl
- php-gettext
- PHPAGI (included with A2Billing)

Monitoring usage

Monitoring your profit margin is essential, and the sample report below shows how easy it is with A2Billing:

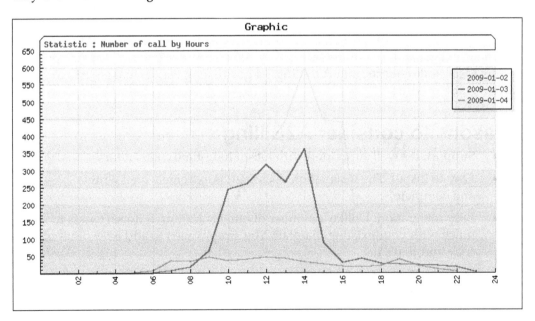

The following table shows the number of calls across the day and the daily costs:

TOTAL							
			ASTERISK MINUTES				
DATE	DURATION	GRAPHIC	CALLS	ACT	TOTAL SELL	TOTAL BUY	TOTAL PROFIT
2009-01-02	2333:09		3139	00:44	88.758 GBP	45.262 GBP	43.496 GBP
2009-01-03	1636:47		1805	00:54	42.472 GBP	14.376 GBP	28.096 GBP
2009-01-04	443:28		467	00:56	11.046 GBP	5.257 GBP	5.789 GBP
TOTAL	4413:24		5411	00.48	142.275 GBP	64.895 GBP	77.381 GBP

Do you think A2Billing is only capable of handling light volumes? Have a look at the following statistics:

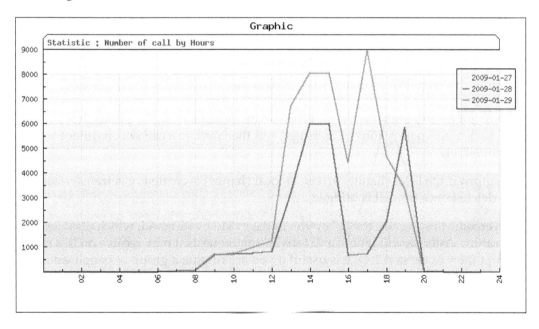

These are "real" calls and not "test" ones, I should add!

Coding for A2Billing

Once you've entered all of your customers, you're reminded that you need to update the config file. Click on update and the customer details for SIP or IAX are written out. It'll even reload the configs for you. Adding the functionality is pretty straightforward. In your extensions.conf file, simply add:

```
[a2billing]
exten => _X.,1,noop(--------------- ${EXTEN} --------)
exten => _X.,2,Wait(1)
exten => _X.,3,DeadAGI,a2billing.php
exten => _X.,4,Wait,2
exten => _X.,5,Hangup
```

It's really that simple—A2Billing allows you to create SIP and IAX entries which are stored as flat files, and, the account code is set to the main account for that customer. When a call is placed, that account code is passed to the PHP script and it bills accordingly. The installation instructions are very comprehensive and lead you through the steps one by one.

Billing gotcha!

There are occasions when Asterisk does not account for calls as one might expect it to.

Consider this scenario—you're hosting a customer that simply has IP phones attached to your server. They discover (on reading the manual) that they can divert calls after a given number of rings. When an inbound call hits the phone, it will send back a "temporary moved message" and then initiate a call which cannot be accounted for.

The culprit is the local channel driver. A local channel is created on a transfer, and then deleted once the call is bridged.

To overcome this, we use the "i" option in the Dial() command, which disables the feature. Asterisk will ignore any forwarding requests it may receive on this dial attempt (new in version 1.4). It is useful if you are ringing a group of people and one person has set their phone to something (for example, forwarded direct to voicemail) that normally prevents any of the other phones from ringing.

High call volumes

A2Billing is reported to be able to handle call volumes of 2.5 million minutes a month and may well scale beyond that, but what about high-call setups? Well there's another project out there called GnuGk (click on `http://www.gnugk.org` to know more).

This solution uses the H.323 protocol (Asterisk supports H.323 in the `addons` package). Reports are that this solution is able to handle at least 30 call setups/ second. You might think that's not a lot, but it truly is. It works out to be over a million calls an hour—big, unless you're a major telco.

What's this H.323? Isn't SIP supposed to be the way to go? Well, not always. Many traditional systems still use H.323 for a number of reasons, not in the least efficiency. The SIP protocol has a large overhead in passing messages compared to H.323.

All these extra messages that get passed to and fro delay call processing as they add extra code to decode them, not to mention all the extra bandwidth. Let's not forget that the messages (as well as the response numeric codes) also transmit rather wordy text explanations as well. If you've ever turned on `SIP DEBUG` in Asterisk, you'll no doubt see huge volumes of text on the console even when your system is quiet.

On a personal note, I fail to see why there's so much bandwidth wasted when no human is likely to read the text. Surely, binary responses would be much more efficient? I'm sure there's a valid historical reason for this (in the early days of the Internet, most data was in the 7-bit format).

Good as this product (GnuGk) appears to be, it's quite clear from the documentation that there is little or no GUI interface. You're going to have to roll one of your own, but that's not always a bad thing.

Other high-call-volume solutions

Are there any other solutions out there? Well, discounting Asterisk-based solutions, there are a couple of solutions that stand out.

SER/OpenSER

For the sake of brevity, we'll refer to the above as SER. OpenSER is a mature fork of SER, which at the time of writing has split into a further fork, but you can read all about that on the net. SER has been around a number of years now and has a large installed base. It can handle large numbers of call setups and registrations. For this reason, many larger setups use SER to frontend multiple Asterisk servers. Why can't you just use SER? Well, SER does not have a **Backend User Agent (BUA)**. Yes, it can handle NAT and message routing extremely well, but if you want to do **Interactive Voice Response (IVR)**, music on hold, and so on, you need a BUA. Asterisk is such a beast. Joining the two together provides a very powerful platform that can scale to thousands of users. However, SER is not for the novice. In order to avoid many hours/days of frustration, it's essential that you gain a good understanding of the SIP protocol. You can always read the RFC #3261 (at `http://www.faqs.org/rfcs/rfc3261.html`.), or you can get books on the subject. If you're looking for a step-by-step guide, we recommend having a look at *"Building Telephony Systems with OpenSER", Flavio E. Goncalves, Packt Publishing.*

FreeSWITCH

What about a high-volume solution that combines the efficiency of OpenSER with many of the features of Asterisk? Is there such a product? Well, actually there is, and it's called FreeSWITCH. One might think of this product as sitting in between OpenSER and Asterisk. Discussion of FreeSWITCH is outside the scope of this book, but it suffices to say that if you are having trouble scaling Asterisk up to very large call volumes, then it may be a viable alternative. Although, care is needed, as FreeSWITCH being a relatively "young" product does not have the same breadth of functionality as Asterisk. As always, you should do your homework first.

Summary

This chapter covers call accounting and how it can be a useful tool in analyzing system utilizations. We examined two billing solutions, which one you decide to implement is up to you. We also looked at high-call-volume solutions. Yes, there are a number of other solutions out there, but the two we explored are known to work. By implementing either of these solutions, you will be able to reliably perform call accounting and bill your customers. But beware—there are many, many "consultants" out there who will try to charge you for a billing solution which will cost you thousands. Be a little bit careful, many will sell you consultancy on these very same projects, which is something you could easily do yourself.

7
Resilience and Stability

In Chapter 2, we touched on one of the issues facing Asterisk in commercial situations, namely, that when implementing Asterisk solutions, you're up against the proven reliability of traditional telephone systems. The big names in the market have invested huge amounts for research and development in perfecting their products, because the first and over-riding expectation of anyone using a telephony system is that each and every time they pick up the phone, they should be able to make or receive a call. As reliability is such an important feature of telephony systems, the tradition has been to tightly control development and release new versions rarely, and after extensive testing. These commercial products are carefully managed, with only a small number of employed developers having access to the source code and developing that code with specific, commercially-oriented goals in mind. Asterisk, meanwhile, is constantly evolving, adding new features, and refining existing ones with both functionality and reliability foremost. It is an open-source product, albeit one "owned" by Digium, where anyone is free to look at the source code and contribute enhancements which they think will improve the product. While this is a well-proven model now, with Linux itself as its shining beacon, it is more difficult to focus the efforts of all potential contributors in specific areas.

Digium's approach has been to concentrate their efforts on a core Asterisk "engine", leaving the myriad of Asterisk contributors around the world to focus on extending that core functionality through add-on modules or products. This is not unlike the approach to Linux itself, where development of the kernel is controlled relatively tightly, but there is plenty of scope for enhancing that pretty basic core, which is the reason for the preponderance of Linux distributions and applications.

The fact that it runs as an application on a wide range of Linux distributions is one of its strengths as well as a weakness. There is no single standard Asterisk install used by one and all, instead there are any number of Asterisk flavors. Some are designed for ease of installation and ongoing maintenance through the use of graphical interfaces, such as the excellent FreePBX frontend. Others eschew the prettiness of a GUI in the name of speed, requiring that all changes be made directly to the text configuration files. Some are set up with scalability in mind, or as billing platforms (which may use A2Billing as the frontend).

In short, whatever be your telephony need, it's highly likely that there is an Asterisk "distro" out there that will cater for it. If there isn't, then you can always create one of your own.

Whichever Asterisk implementation you choose, it's easy to come to the righteous belief that you're going to change the world with it, using its wondrous features and low cost. But if the customer can't reliably make calls, then all the features in the world aren't going to placate them.

This chapter will endeavor to give you some techniques on how to make your Asterisk implementations both resilient and highly stable. While some of the techniques require you to edit the configuration files directly, you will have no trouble doing this even if you have implemented a version of Asterisk with a GUI frontend, such as FreePBX.

Increasing availability

Making a system more resilient requires a two-pronged approach. Before examining this, it is worth remembering that the goal of this exercise, as always, should be considered from the customer's point of view. In this case, the goal is to maintain the customer's telephony operations. Obviously each customer will have constraints, usually budgetary, which restrict the extent of the measures that can be taken to achieve this goal. But transparency and an inclusive decision making process will avoid much recrimination later if, or rather, when, something goes wrong. Because, if you wait long enough, something will always go wrong.

With any operational system, the ideal is to have 100% uptime. However, achieving this ideal is likely to be beyond the budget of many organizations, as it relies on the use of an extensively tested and completely unchanging hardware and software combination. In such a combination, each component has been constructed for a precise purpose, and to hugely-exacting standards. Given that every Asterisk system has, at its heart, standard PC architecture, the 100% uptime condition clearly isn't going to happen even with the best will in the world.

Mentioning standard PC architecture to the average customer runs the risk that the PC under their desk that crashes on a regular basis will spring to mind, so it's a phrase best used with caution. Of course, much of the world's IT infrastructure now runs on standard PC architecture, including the systems that keep hospitals operational and governments running. It is how these server infrastructures are constructed that provides the lessons which our relatively speaking, modest telephony system can use to keep the customer's phones ringing.

This brings us back to our two-pronged approach. In attempting to maximize uptime of any system, it is sensible to do all that is reasonable to make the system as reliable as possible. But you should also assume that the odds are against achieving 100% uptime indefinitely. Therefore, it is also sensible for critical systems to have a contingency plan in place—a means of maintaining operations if a part of the system fails. This type of resilience is called **failover**. Let's look at means of boosting both stability and resilience in turn.

Stability

A telephony system consists of a number of parts, each of which needs to be looked at when considering stability. A typical installation will have most or all of the following components:

- Network (cables, routers, switches, and so on)
- Endpoints (telephones)
- Telecom switches or gateways (PRI/BRI/Zaptel[or DAHDI]/GSM)
- Server (PBX)
- Environment

Any IT system, like a chain, is only as stable as its weakest link. If you wish to boost the stability of your system as a whole, then you need to consider how you can improve each aspect of the system.

Network

In all likelihood, some or all of the network components used by the new telephony system will be shared with the data network. Because of this, it's possible that the elements of resilience have already been introduced to boost stability. If not, implementing a new PBX over the existing network infrastructure is the ideal opportunity to ensure that both voice and data traffic traverse a path which is "fit for purpose". Usually that means introducing redundant equipment, or components to the network with the purpose of "stepping in" in case of failure. However, there are ways in which you can improve the individual aspects of your network, such as cables, switches and routers, so that they are likely to be the most stable parts of your telephone system.

Cables

Cabling is the simplest component. But with a large installation, you may need to purchase significant quantities for both patching in the communications room/area, and for connecting the endpoint devices to network points in the workplace, making it tempting to shave pennies off the cost per meter. They tend to be very reliable, although opting for the cheapest available is likely to be a false economy. It is not unusual for cabling to be outsourced to a specialist company. If you're comfortable making your own cables, then there is a lot to be said for that approach too. Investing in CAT-5e, or even CAT-6 cables is sensible, particularly if gigabit switches are planned to be used. It is also sensible to ensure that a color scheme is agreed with your customer so that telephony cables can be easily differentiated from data cables.

Switches and routers

Looking at switches and routers, it is usually in the customer's long-term interests to purchase a named brand, as the vendor will have carried out appropriate testing and development to ensure that their products are likely to work without fault for many years, and will provide warranties to cover failure in that time. Leading enterprise providers include Cisco, Foundry, Extreme, and Juniper, with the likes of HP and Dell also having some share at this end of the market. For smaller installations, a good cost/feature balance can be achieved with HP, Dell, 3Com, and even top-end consumer equipment from Netgear and D-Link.

It can be tempting for the customer to save money by specifying Layer 2 switches, but if there is any prospect of using VLANs and/or QoS in the future, then you should strongly recommend that Layer 3 switches are used from the start. It is also worth having a conversation with the customer about **Power over Ethernet (PoE)**, at least for the endpoints that would be expected to work in the event of a power cut.

Endpoints

Similar cost/benefit principles also apply for the choice of endpoints. It is preferable to spend an extra 10% or 20% on devices that have a good quality reputation rather than trying to save some pennies now, only for your customer's business to be adversely affected some time later. This may mean choosing the likes of Polycom, Aastra, Snom, or even Cisco, now that they support SIP, over a cheap, no-name brand. Indeed, the choice of endpoint is even more significant as it is likely to be the only physical contact most users have with your system, and perceived quality will be an important factor in the ongoing acceptance of the system.

Telephony switches and gateways

A telephony switch will merely guide the telephony traffic within your network, usually on the basis of either a destination IP address or MAC address. By definition, this needs to be separate from the terminating device(s), so figuratively speaking, it will normally appear in your installation as a little black box. It is usual for telecoms switches to be used in conjunction with PBXs containing internal termination cards.

Telephony gateways convert telephony traffic from one form to another. This can happen either within the PBX, in the form of a plugin PCI card that passes the telephony voice and data to Asterisk running on the host processor, or it can happen in an external device. Using a separate telephony gateway device, while important for failover scenarios as we will explore shortly, is also preferable purely from a system resilience point of view. Keeping our server as simple as possible is always a laudable goal, so putting a termination card in there is going to increase the risk of failure. Having the termination happen in its own purpose-built chassis, with a well-matched power supply, and appropriate cooling measures, will also maximize its likelihood of achieving the required uptime.

Deciding which approach is more appropriate for your installation is normally only a decision that is relevant for smaller systems, as terminating multiple PRI lines in a single, large PBX is a tough decision to justify from a stability/resilience viewpoint. However, for a small installation with either a BRI or even a couple of analog lines, keeping as much as possible in a single, wall-mounted box may be preferable to finding some floor space in a busy office, for a half-height cabinet. As long as the customer understands the compromises involved, it's a valid decision to take.

Asterisk-friendly switch and gateway vendors include Xorcom, Redfone, Junghanns, and Rhino.

Server

With the server, we will be using standard PC architecture. However, this doesn't mean that any old PC off the shelf should be considered. Most telephony systems are required to be working 24/7/365, so you should approach your choice of hardware with that in mind. In all honesty, the easiest solution is to purchase a purpose-built server from one of the established vendors in this arena, namely the likes of HP, Dell, and IBM. Their 1U and 2U server offerings are usually powerful enough for all but the largest installations. They are very cost-effective and in our experience, are highly stable too. In scenarios where a rack-mounting is not viable, typically for the smaller installation, there is usually a smaller floor-standing option from these vendors that contains essentially the same components.

When choosing an appropriate server, you should seriously consider the following features:

- Dual power supply
- Mirrored system HDDs (RAID 1)
- Striped data drives (RAID 5)
- Dual network cards
- >1 memory module
- Dual processors

Implementing all of the above can push the price up considerably over a basic server, so this should be discussed with the customer. Certainly, processors tend to be the most stable of the components we listed, so any compromise there is unlikely to seriously increase the risk of failure.

If a highly-resilient single server is required, then the HP DL360 range, as shown, provides a good price/performance ratio.

Environment

Of course, spending money on good quality, highly-resilient components can be undermined if they are not used properly or are located in an environment likely to cause failure. For instance, a server or switch is more likely to fail if placed unprotected in a frequently used broom cupboard. Most cables will stop working pretty quickly if left unprotected on the floor for people to walk on and chairs to roll over. If your customer does not already have a secure and ventilated area for their servers and communications equipment, then your proposal must recommend that one be provided. Even a wall-mounted small communications cabinet is adequate and inexpensive.

In addition, if any telephony is likely to be required if the power fails, then a UPS is absolutely essential. As already mentioned, not all endpoints need to be connected to a PoE switch, just the ones that are deemed necessary during a blackout. In addition, you should make sure that all the other elements of the telephony system are also protected, including the server, PoE switches, backbone switches, and gateways. Bear in mind that you may need an extra UPS if the telephone lines enter the building at a point other than the server/communications room.

Dealing with failure

When specifying the hardware for a new Asterisk installation, it's tempting to buy top of the range fully-loaded servers. However, when considering ways of maximizing uptime, lots of lower-spec servers are better than one big one. It doesn't matter how many redundant components you have in a well-specified server, there's always a single point of failure somewhere. To illustrate, a couple of years ago, a customer insisted on the most powerful/redundant server available. The server had dual processors, dual power supplies, lots of fans, and a big RAID array. However, despite that wonderful spec, it went offline for two days recently. The cause of the failure? A dry joint in the power button circuit.

The commonly accepted approach to maximizing system availability is to assume that, at some stage, you are virtually guaranteed to experience the failure of a component. In order to cope with that failure, you need to be able to recognize it has happened, and use an alternative component to ensure that no system downtime is experienced, that is to say, to "failover". The problem is that in telephony systems and data networks, there are probably hundreds, if not thousands, of components to consider. The ultimate goal is to have no single point of failure (a component with no backup). But usually, the elements of the system least likely to fail are left unprotected, as it is not cost-effective to introduce redundancy. To start with, let's look at how to make a data network more resilient.

Network resilience

Network design takes the standard approach to resilience, namely to assume that certain components may fail, and to allow for that fact by running two or more components in parallel. Resilient routers and switches, for example, will utilize dual power supplies just in case one of them fails, and may also have dual processors for the same reason. Such devices cost more than non-resilient ones, but will provide seamless operation if covered components fail, if that is important to your customer.

 Usually, a calculation of the potential disruption and lost revenue makes a compelling case for a slight increase in spend on resilient network equipment.

While using devices with an appropriate level of resilience is important, a truly resilient network will assume that devices will eventually fail no matter how resilient you make them. The usual mechanism for implementing such a failover situation is to have primary and secondary devices connected together, with the devices sending regular signals to each other to let them know that all is well. For obvious reasons, this is known as a **heartbeat**.

The two devices should also synchronize their configuration constantly. If the heartbeat fails, the secondary device assumes responsibility for the traffic. It is not unusual for some user notification mechanism (for example, an email) to be triggered when this occurs, as the failover should be so seamless that it is not humanly detectable. Of course, not all devices are capable of failing-over in this fashion, so you need to know what facilities you require when you make your initial purchase.

Improving network resilience, or its ability to deal with failure of nodes, can often mean compromising the lean approach that is taken when designing a network with speed in mind. In order to make the improvement, alternative routes need to be in place when the primary route fails, whether that failure is due to faulty devices or cables. Those alternative routes need to span both the LAN and the WAN, as the most resilient internal network in the world is no good if there is a single internet circuit that fails. It is worth bearing in mind that the circuit failure may be due to an issue at an ISP, or due to something physical that is much closer to home, such as a digger going through a cable. Utilizing both logical and physical separation in redundant circuits is wise. However, once multiple circuits are in place, there is no reason why any circuit should remain idle waiting for a failure elsewhere. Rather it makes sense to ensure that adequate bandwidth is allocated to voice traffic when recovering from a fault, probably through the use of QoS.

The failure of a switch can be catered for in a slightly different way, assuming that you have three or more switches in use. Instead of having switches paired up, so that you need twice the number that a "lean" configuration demands, you can simply connect each switch to two others, such that, if one uplink cable or even a switch fails, there is always a route through to the other active switches. The devices connected to the failed switch will require re-cabling to an active device. However, traffic up and down the backbone will not be stopped. Critical servers, of course, will have at least two network cards that should be cabled to different switches, to ensure that a switch failure does not cause a loss of service. The following illustration shows a backbone switch cabling strategy:

The potential problem with cabling switches in this way is that routing loops may be introduced. But this has been catered for by the use of the **Spanning Tree Protocol (STP)**, which creates a map of possible routes from one node to another and weighs them according to criteria such as number of hops and connection speed. Not all switch equipment is Spanning-Tree capable, so the old adage of "let the buyer beware" applies. It is also recommended by Microsoft that switch ports with clients connected to them should have spanning tree disabled to avoid potential issues with DHCP, as stated at: http://support.microsoft.com/kb/q168455/

Re-routing traffic, somewhat surprisingly, can be more difficult to achieve effectively, if you wish to cater for cable issues anywhere on the network. However, while the loss of communications on your network backbone (between two switches, or between a switch and a router) can affect many people, the loss of a cable between a phone and floor port should only affect one. Therefore, in most cases it is appropriate to have different strategies for these different situations, most likely accepting that a cable issue that disables a phone, can be quickly and easily dealt with by replacing the cable.

Server

As we have already seen, it's fair to assume that there is a single point of failure somewhere even in the best-specified server. Bearing this in mind, the means of assuring continuity of service is to have another server ready to take over. As Asterisk is Linux-based, one means of achieving this is to use a Linux project called Linux-HA.

High availability

The Linux-HA project, in its own words, aims to:

> *Provide a high availability (clustering) solution for Linux which promotes reliability, availability, and serviceability (RAS) through a community development effort.*

In Asterisk systems, the most widely used product based on Linux-HA is Ultra Monkey, which is often used in conjunction with commercial hardware that allows communications circuits to be switched between PBXs. For medium to large installations, both Redfone and Junghanns produce fault-tolerant PRI gateway/switch products specifically aimed at Asterisk. This setup will normally be used within an active-passive failover scenario, although there is no reason it cannot be used for an active-active scenario to introduce some load distribution.

Ultra Monkey

Ultra Monkey is an open source project that provides software for Linux installations to enable them to be configured in a high availability cluster. The software sets up a virtual IP address for two or more Linux servers. It also sets up the service (as discussed earlier) known as a "heartbeat"—a means of determining if a server in the cluster is still active. The heartbeat requires that the servers have a means of communicating separate from the main LAN. This is normally achieved through use of a second NIC in each server connected to a separate switch, or to a separate VLAN. Installation instructions vary depending on which Linux distribution your Asterisk system is built on, but some popular distributions are covered at `http://www.ultramonkey.org/3/installation.html`.

In the next figure, we see a typical use of Ultra Monkey to provide a highly-available-active/passive Asterisk system. There are two Asterisk servers, primary and secondary. There is one main LAN carrying all the voice traffic and a secondary LAN for heartbeat traffic, which could also be used to carry the mirroring/synchronization traffic. There is a failover device on the network interfacing with the ISDN circuits, allowing this function to be physically separated from the Asterisk servers. In normal use, the primary server handles all call traffic, which can be a combination of VoIP and PSTN calls. The main LAN uses addresses in the range 10.1.x.x and the heartbeat LAN/VLAN uses addresses in the range 10.2.x.x.

An alternative to the dual-LAN mechanism is simply to connect the servers together using a serial cable, which works perfectly well if you only have two servers.

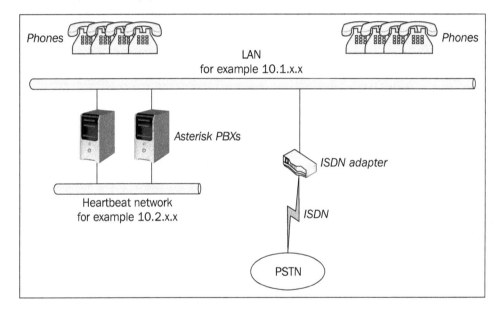

Let's assume the primary server above has the static IP address 10.1.0.1 on the main LAN and the secondary server has the address 10.1.0.2. A floating IP address, say 10.1.0.3, is associated with one of these servers by Ultra Monkey. The heartbeat travels around the 10.2.x.x VLAN, ensuring that all nodes are available, in which case the floating IP is associated with the primary server. If the primary server is lost, then the software can run predetermined scripts and then switch the floating IP to the secondary server. All other devices, such as phones, can register with the floating IP in preference to the static IP.

Telephony switches/gateways

In the previous illustration, within our high availability system, the external telephony circuits are terminated by a gateway device. This then directs the traffic at the currently active Asterisk server. An alternative would be to have a PCI card in each PBX, and a telephony switch directing the ISDN traffic at either the primary or secondary PBX. Without a switch or gateway, the failure of the primary PBX would either require the termination card to be transferred as a whole to the backup PBX, or at least to have the cable moved from one PBX to the other if both had identical cards. Either way, the time taken to recover from a failure situation has suddenly risen from seconds to minutes, or even hours.

The manufacturers of the two devices we are going to look at have made a point of targeting the Asterisk market, and their products can fairly be described as being mid-range. Even though they work in slightly different ways, much of the call handling is still carried out by the PBX. This means that registration needs to be with a physical server rather than a floating IP address, adding a little extra complexity to the failover process. However, they provide a very cost-effective means of building a high availability system.

An alternative approach would be to use a true ISDN-SIP gateway which presents ISDN channels as SIP trunks to the Asterisk PBX. With this setup, the SIP trunks are registered to the floating IP, and at failover, the secondary server would simply re-register them. This approach is easier, but much more expensive as the gateway carries out the line terminations.

Redfone foneBRIDGE2

The **Redfone foneBRIDGE** is a device that will actually terminate a telephony circuit (for example, a PRI), and direct the traffic to a PBX using **TDMoE** (**Time Division Multiplexing over Ethernet**). Thus, the Asterisk servers do not require ISDN termination cards, and can simply use zaptel (or DAHDI) to deal with the incoming call traffic.

 It is recommended that you have a physical zaptel (DAHDI) interface somewhere on the network, or use the `ztdummy` (`dahdi_dummy`) kernel module, to provide the appropriate timing information.

With a very simple script, the device can be reconfigured on the fly to direct traffic to a different server. Therefore, in a failover situation, heartbeat can run a script to register the foneBRIDGE to the secondary server, another script to start Asterisk on the same server, and your failover is complete, all within thirty seconds.

As the device pumps out TDMoE, which can be routed over a LAN (but not the Internet), the foneBRIDGE2 can be used to manage failover between many Asterisk servers.

Junghanns ISDNguard

The Junghanns ISDNguard is an ISDN switch, and thus requires any of the connected Asterisk servers to be able to terminate the ISDN traffic themselves. It has its own means of determining if the primary Asterisk server has failed, by using a serial cable connection. If it loses communication with the primary server, it switches the ISDN line(s) over to the secondary server. Although this device can be used in an active-passive failover scenario, it would also require the use of Ultra Monkey or other Linux-HA software to manage the provisioning of the passive server. However, it is ideally suited to an active-active failover scenario, where the secondary server is ready and waiting for the switch. As there needs to be a direct connection to the termination cards in the servers, this device will only work in a two-server failover scenario.

The ISDNguard is a simpler device as compared to the foneBRIDGE2, and that is reflected in the relative cost. However, in order to implement a failover system with the ISDNguard, two interface cards are also required. Using Digium TE210P dual T1/E1 cards will bring the total cost to around the same as the foneBRIDGE2.

Endpoints

One issue with failover, even in an active-active setup, is how to manage calls that are active at the time of failure. In principle, it should be possible for SIP calls to carry on regardless. This is because the Asterisk server is, in theory, only required to set up the call in the first place and to manage such features as music on hold, or call recording if they are required during the course of the call. This behavior is controlled using the `canreinvite` = yes|no option in `sip.conf`, and is described in good detail at the following URL:

```
http://www.voip-info.org/wiki/view/Asterisk+sip+canreinvite
```

In a nutshell, SIP devices set up a communication channel between themselves if, and only if, both devices are capable of supporting this type of call and of allowing Asterisk to reinsert itself into the call channel, if any of the server side features we saw just now are required. In SIP terms, this is called a **re-invite**. Obviously, the choice of endpoint, both on your side and the call recipient, is crucial. So even if you have everything configured just right in your system to allow for calls to persist during a failover, you may find that some still drop because the other side isn't quite there yet.

Of course, another major barrier that tends to make persistent calls difficult to achieve is the use of NAT. If either or both ends of a call are behind NAT firewalls, then setting up a dedicated communication channel is only possible if a SIP proxy is used, which adds a whole new layer of complexity into the mix.

Another issue with endpoints during failover, is the need for them to re-register with the secondary server. As we saw in Chapter 2, especially in large installations, it is usually better for endpoints to ping the server to maintain registration, and if this is possible, then pinging the floating IP address should cause the endpoint to register with the secondary server. Of course, if your endpoints only do this every sixty seconds, then it could be that long before re-registration occurs.

It can take an awful lot of work to achieve the nirvana of persistent calls during a failover, which your customer may baulk at funding, given that the risk of failover occurring in the first place should be quite miniscule.

Round robin DNS

So how do you manage the load on your servers? One option is to use round robin DNS, which is a feature of many DNS servers (open source and commercial). Instead of constantly giving the same IP address out for a domain name request, the device holds a list of IP addresses. As each request comes in, either the address at the top of the list is handed out, or more frequently, the complete list is offered. Then the address at the top is moved to the bottom of the list. The next request results in the next IP address being handed out, either by itself, or at the top of the list, and so on. Eventually, the first address comes back to the top of the list.

Strictly speaking, round robin DNS is not a load balancing technique, but rather a load distribution technique. There are other techniques available that more accurately divide calls, based on how heavily individual servers in a cluster are loaded. Quite often, these require that a specialized router intercepts inbound traffic, and distributes call load appropriately.

Round robin DNS is typically used for geographically distributed servers, where you want to ensure that there is a relatively equal spread of load and that, if one location becomes unavailable, the others can still service requests (the device that passes out IP addresses should be constantly checking that the service at each address is available, and should remove it from the list for those not available). There are a couple of potential issues with this technique, such as:

1. Most ISPs set this to twenty four hours, although, there are a number of DNS service providers that will allow it down to five minutes. If the TTL (Time To Live) is too high, then the next request from the same external device will have the previous results cached, so it will not pull the new IP address(es) from the DNS server.

2. There is an assumption that all requests will result in a similar amount of work for each server. However, this may not be the case for Asterisk installations, where one incoming call may last for twenty seconds, and another could last for two hours. If one server receives a disproportionate amount of the longer calls, then it could have a much higher load than the other servers.

    This technique will not work for outbound calls, if the phones in use have the ability to register only with a single Asterisk server. Therefore, for this, as with most Asterisk failover techniques, it is wise to use phones that can register with more than one server, ideally failing over to a second server if the first is unavailable. The good news is that this functionality is a standard feature of most business IP phones.

Say hello to Rsync

Assuming you have successfully implemented load balancing/distribution using round robin DNS, or a failover setup, perhaps using Ultra Monkey, the next issue is to synchronize the server dial plans and `iax/sip.conf` entries across your servers. After all, you really want to make changes only on one server (master), and have it propagate to the others. There are products that will do this for you. In the open source world, DRBD is quite popular as a byte level mirroring tool, but there are simpler solutions too.

To synchronize the various `.conf` files, you can use the excellent Rsync program, included by default in most Linux distributions. Rsync is a program that synchronizes files between systems. What makes it special is that rather than copying an entire file each time it changes, instead, it copies the "changes" from the source system to the destination. If no change occurs then nothing is copied across. Rsync can keep individual files, directories, or even whole file systems synchronized, and can optionally preserve symbolic links, hard links, file ownership, permissions, devices, and times. It can pipeline changes to multiple files to increase speed, and use `rsh`, `ssh`, or direct sockets as the transport mechanism between trusted servers.

As Rsync only copies changes that have occurred since the last synchronization, there is little overhead in having many syncs per day rather than very few. From a failover perspective, of course, it's better to have the changes passed round the estate as soon as possible, as failure is rarely so accommodating as to occur immediately after your once-a-day synchronization job.

In essence, what we're going to do is set up a cron job to copy any changes in the master to all the slaves. So, do we just need to copy all the Asterisk directory files? Well, not quite. Some files, or at least sections of some files, are going to be machine specific, such as `iax.conf` and `sip.conf`. What we need to do here is create stub files, and `#include` the main configuration files. We then use the `exclude` option in Rsync to exclude the stub files, which contain the machine specific code. The following code snippet shows a stub `sip.conf`:

```
[general]
bindport=5060; UDP Port to bind to (SIP standard port is 5060)
localnet=192.168.10.0/255.255.255.0; All RFC 1918 addresses are local
networks
#include sip_additional.conf
```

This file, being machine specific, will be in the exclude list, but `sip_additional.conf` will be propagated across the network. If you are planning to run more than one secondary server, then it is wise to ensure that all the secondary servers do not pull their configuration from the primary server. The reason being that in the case of primary server failure, your secondary servers will run the risk of gradually going out of sync until the primary server is returned to service (they will have no "source" for the sync job). Instead, it makes more sense to sync each server with two others, being a sync job, it will push changes both ways.

Running this as a cron job every five minutes won't put an undue load on your servers and will ensure they are synchronized. Reloading Asterisk's config completes the picture.

The first cron job syncs the `/etc/asterisk` directory with an Asterisk server at the IP address—`192.168.10.138`, but excludes five files which contain machine specific information. This first cron job involves the following code:

```
/usr/bin/rsync -v -t  --exclude=sip.conf
--exclude=a2billing.conf --exclude=manager.conf
--exclude=iax.conf   --exclude=extensions.conf
--progress --partial rsync://asterisk@192.168.10.138
/asterisk/*  /etc/asterisk >/dev/null 2>&1
```

The next cron job duplicates the voicemail directory with the same remote server, applying the code:

```
/usr/bin/rsync -aPvt  --progress rsync://asterisk@192.168.10.138/ast_
VM/./var
/spool/asterisk/voicemail/default >/dev/null 2>&1
```

The final cron job reloads the Asterisk's config with the code:

```
/usr/sbin/asterisk -rx 'reload' > /dev/null 2> /dev/null
```

With multiple secondary servers, a little care should be taken over the timing of the cron jobs to ensure that a logical sequence is preserved. For example, with a six-minute cycle between a primary and two secondary servers, you might sequence the servers as follows:

- 0 minute: Sync primary and secondary 1
- 2 minutes: Sync primary and secondary 2
- 4 minutes: Sync secondary 1 and secondary 2

Thus, all servers are synchronized with each other, and there will be plenty of time for each job to be completed. In fact, as already stated, the jobs could be run closer together and, as only changes are pushed across, there should still be no problems. However, as the time between the jobs gets shorter, it becomes even more important that the server clocks are also in sync. So if you have not done so already, now is the time to use NTP to set the server clock accurately, either with an internal NTP server or with external ones such as the `ntp.org` servers, making sure that you use servers local to you.

Limiting the number of calls per server

When looking to balance the load over a number of Asterisk servers, there are several little publicized options in `asterisk.conf` as shown in the following code snippet, which are very useful:

```
[options]
maxload = 0.9    ; The The maximum load average we accept calls for per
processor core (if  dual  core, set value to 1.8)
maxcalls = 140   ; The maximum number of concurrent calls you want
; to allow
timestamp = yes
```

The options are self-explanatory, but you should be aware that the `maxload` and `maxcalls` options will reject calls if the values set are **exceeded**.

The `timestamp` option is also useful as it timestamps the command line interface, which is great for debugging.

Summary

Reliability, in Asterisk terms, refers to the reliability of the telephony system, including the physical server, interface devices, and the network. Individual servers can improve reliability through the use of redundant parts, however, it is difficult to remove all single points of failure. A more reliable setup is to have multiple, smaller servers working together to spread the load and take over if one of them fails.

An effective, high availability solution can be implemented using Ultra Monkey and failover telephony switches or gateways. Multiple servers can easily be kept in sync using the Rsync program included in most Linux distributions. One technique for spreading the load of inbound calls is to use round robin DNS, which works without much intelligence, simply allocating each call to the next server regardless of load.

8
Localization and Practical Security

Have you ever considered the changes that would need to be made if you wanted to put your telephone system in a different country? If you have, you either have too much time on your hands (and you need to chill out a bit more!), or you are a system manufacturer thinking about supplying your product outside of your home market.

Although it is easy to tell that Asterisk was born in the USA, both Mark Spencer and the numerous contributors to Asterisk have kept internationalization (to them) or localization (to us) in mind, as can be seen in their designs.

If your customers are used to hearing a certain dial tone when they pick up a phone, anything other than that tone will be unsettling to them, and that goes for a multiplicity of other characteristics of a telephony system too.

All sorts of things need to be considered, from local numbering plans, to the call progress tones, which are recognized and generated by the system, to the language (and accent) of the prompts, to the way numbers are formatted and spoken, to the physical and electrical characteristics of the telephony interfaces, and some stuff in between.

If you want to implement Asterisk-based solutions outside of North America, then you will need to do some localization my friend!

In addition to localization, we will take a quick look at some practical steps which you can take to improve the security of your Asterisk-based solutions. Needless to say, the main security shield for any given installation will be at the "whole network" level. However, if we are going to allow callers to get "into" our Asterisk, then we would be wise to enhance the level of security where possible.

In this chapter, we will cover the following:

- How to change the tones generated and recognized by Asterisk
- Where and how to affect the way times, dates, and so on are announced
- How to change the language of system prompts
- What needs to be changed in terms of telephony interfaces
- Where to change the method of caller ID signaling used
- A checklist to ensure you have localized everything you should
- Some ways to secure your Asterisk against unauthorized use by internal and external callers

Tones

Let's work from the inside out. The types of call progress tones played to the callers once they are within Asterisk are set in indications.conf, which is, of course, one of the many .conf files to be found in the /etc/asterisk directory.

On opening indications.conf, you will find that the default set of tones to be used (by the pbx_indications module) is specified in the [general] section with a two letter country code.

The next example, shows the way the first few lines of the file look on a fresh installation, prior to any changes:

```
; indications.conf
; Configuration file for location specific tone indications
; used by the pbx_indications module.
;
; NOTE:
;     When adding countries to this file, please keep them in
alphabetical
;     order according to the 2-character country codes!
;
; The [general] category is for certain global variables.
; All other categories are interpreted as location specific
indications
;
;
[general]
country=us                    ; default location
```

The highlighted code shows the country specified by the two letter country code. If `indications.conf` is missing, Asterisk will assume that you want to use the US tone set.

There are a number of separate sections further down in the files, each headed with a two letter country code in square brackets, which describe the actual tones for that country in terms of frequencies and cadences, along with the odd explanation or cryptic note left there by the compiler of the tones for a given country.

Here we see the entry for the UK:

```
[uk]
description = United Kingdom
ringcadence = 400,200,400,2000
; These are the official tones taken from BT SIN350. The actual tones
; used by BT include some volume differences so sound slightly
; different from Asterisk-generated ones.
dial = 350+440
; Special dial is the intermittent dial tone heard when, for example,
; you have a divert active on the line
specialdial = 350+440/750,440/750
; Busy is also called "Engaged"
busy = 400/375,0/375
; "Congestion" is the Beep-bip engaged tone
congestion = 400/400,0/350,400/225,0/525
; "Special Congestion" is not used by BT very often if at all
specialcongestion = 400/200,1004/300
unobtainable = 400
ring = 400+450/400,0/200,400+450/400,0/2000
callwaiting = 400/100,0/4000
; BT seem to use "Special Call Waiting" rather than just "Call
Waiting" tones
specialcallwaiting = 400/250,0/250,400/250,0/250,400/250,0/5000
; "Pips" used by BT on payphones. (Sounds wrong, but this is what BT
;   claim it
; is and I've not used a payphone for years)
creditexpired = 400/125,0/125
; These two are used to confirm/reject service requests on exchanges
;   that don't do voice announcements.
confirm = 1400
switching = 400/200,0/400,400/2000,0/400
; This is the three rising tones Doo-dah-dee "Special Information
;   Tone",
```

```
; usually followed by the BT woman saying an appropriate message.
info = 950/330,0/15,1400/330,0/15,1800/330,0/1000
; Not listed in SIN350
record = 1400/500,0/60000
stutter = 350+440/750,440/750
```

So, once you have chosen your country by changing (if necessary) the two letter country code, you just need to save the file and reload.

You will now hear those tones when a call is being handled **inside** Asterisk. The word "inside" has been stressed as all calls are initiated outside of Asterisk, and will come in from SIP, IAX2, or analog phones, or trunks of one description or another. The nature of the tones you hear through these devices is not the responsibility of Asterisk, although Asterisk will dictate the type (busy, ringing, and so on) that you hear. Therefore, now we need to step outside of Asterisk and deal with those tones.

SIP and IAX phones are easy to deal with, as these devices generate the tones themselves. So, when you lift the handset on say, an SIP phone, you will hear a dial tone whether the device is connected to an IP telephony network or not.

Contrast this scenario to the original reason for the dial tone. The clue is in the name "dial tone". It was originally a "confidence tone" to signal to the users that they were indeed connected to a telephone exchange, and that they could now dial a number. Of course, this is still the case with traditional telephony connections. How things changed in the world of IP—all a dial tone in the earpiece of an IP phone will tell you is that your IP phone is alive, and the curly cord to the handset works.

In order to change the nature of the tones you hear from your SIP or IAX phones, you will need to change parameters on the phones themselves, usually, via their web interfaces. Asterisk only tells these telephones which tones to play (through the protocol being used for call control). It is on the phones themselves that you will need to change the tones.

We have now seen how to change the tones that Asterisk provides to calls which it is terminating, and we have seen that the tones heard through SIP and IAX phones (prior to the call being terminated by Asterisk) must be changed on the devices themselves.

Let us now consider the tones that will be used (both recognized and generated) for analog telephony. If analog devices are connected through **ATAs** (**Analog Telephony Adaptor**), then the tones will again be changed on the ATA devices themselves, as tone generation and recognition is part of their job.

However, if we connect our analog devices to Asterisk through a Digium card, we will need to configure these tones as part of the DAHDI (formerly Zaptel) setup.

In the `/etc/dahdi/` directory a file called `system.conf` (formerly `etc/zaptel.conf`) will be found with the code:

```
loadzone=us
loadzone=uk
loadzone=nl
defaultzone=us
```

We are interested in two parts of the configuration in this file — `loadzone` that loads a set of tones to be used with the analog card(s), you can load as many sets as you want (details of the actual tones are found in `zonedata.c`), and, `defaultzone` that defines the tone set you will use as standard when handling calls.

Remember, we are only talking about the tones that will be recognized or generated on analog channels here.

If you want to employ some tones which are loaded, but not default, this is easily done within the dialplan using the `Playtones()` application.

That was about the tones. Tones that will be heard once a call is inside Asterisk, such as the following, are chosen and specified in `indications.conf`.

- Tones heard on SIP or IAX phones (prior to the call being answered by Asterisk) are set on the devices themselves

- Tones to be generated or recognized on analog channels through Digium cards are loaded (and defaults set) in `etc/dahdi.conf`.

Take a look at the next figure to reinforce this:

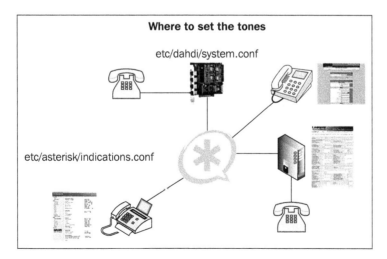

Time and date and localization

Although the actual system time and date for your Asterisk machine will only be affected by the setting in Linux, it is quite possible to make Asterisk aware of as many time zones as necessary, and to alter the way time and date are set out and spoken within Asterisk.

When you think about the need to manifest different time zones and spoken localization (other than the language itself — we'll come to that), it really occurs only in voicemail transactions.

When you are being told at what time and on which date a message was left, you will want to hear it in your local time zone, regardless of where the Asterisk server is located. Moreover, you will want to hear the time and date spoken in a way that you naturally understand, rather than having to struggle to interpret the information while it is given to you in some alien form.

Fortunately, Asterisk recognizes this fact and allows different "voicemail zones" to be set, which dictate the time zone and the way in which the time and date are announced. These settings are created in the [zonemessages] section of voicemail.conf, and further down the file, the voicemail zone for each voicemail box can be specified as an option (if required) where voicemail boxes are specified.

Here is a look inside voicemail.conf to see what we have been talking about. This example starts with comments (identified by a leading semicolon — ";"), which explain each of the options that could be used to construct a voicemail zone. This is followed by the [zonemessages] section where each of the zones is set, and finally we see the use of two of the defined zones (Eastern and European), by being set as options, in the mailbox definitions at the end of the file in the following example:

```
; Users may be located in different timezones, or may have different
; message announcements for their introductory message when they enter
; the voicemail system. Set the message and the timezone each user
; hears here. Set the user into one of these zones with the tz=
;   attribute in the options field of the mailbox. Of course, language
;   substitution still applies here so you may have several directory
;   trees that have alternate language choices.
;
; Look in /usr/share/zoneinfo/ for names of timezones.
; Look at the manual page for strftime for a quick tutorial on how the
; variable substitution is done on the values below.
;
; Supported values:
; 'filename'    filename of a soundfile (single ticks around the
;               filename required)
; ${VAR}        variable substitution
```

```
; A or a        Day of week (Saturday, Sunday, ...)
; B or b or h   Month name (January, February, ...)
; d or e        numeric day of month (first, second, ...,
;               thirty-first)
; Y             Year
; I or l        Hour, 12 hour clock
; H             Hour, 24 hour clock (single digit hours preceded by
;               "oh")
; k             Hour, 24 hour clock  (single digit hours NOT preceded
;               by "oh")
; M             Minute, with 00 pronounced as "o'clock"
; N             Minute, with 00 pronounced as "hundred" (US military
;               time)
; P or p        AM or PM
; Q             "today", "yesterday" or ABdY
;               (*note: not standard strftime value)
; q             "" (for today), "yesterday", weekday, or ABdY
;               (*note: not standard strftime value)
; R             24 hour time, including minute
;
;
;
; Each mailbox is listed in the form <mailbox>=<password>,<name>,
; <email>,<pager_email>,<options> if the e-mail is specified, a
; message will be sent when a message is received, to the given
; mailbox. If pager is specified, a message will be sent there as
; well. If the password is prefixed by '-', then it is considered to
; be unchangeable.
;
; Advanced options example is extension 4069
; NOTE: All options can be expressed globally in the general section,
; and overridden in the per-mailbox settings, unless listed otherwise.
;
; tz=central Timezone from zonemessages below. Irrelevant if
; envelope=no.
; attach=yes            ; Attach the voicemail to the notification
; email *NOT* the pager email
; attachfmt=wav49       ; Which format to attach to the email.
; Normally this is the
; first format specified in the format parameter above, but this
; option lets you customize the format sent to particular mailboxes.
; Useful if Windows users want wav49, but Linux users want gsm.
; [per-mailbox only]
; saycid=yes            ; Say the caller id information before the
;     message. If not described, or set to no, it will be in the
;     envelope
```

```
; cidinternalcontexts=intern   ; Internal Context for Name Playback
; instead of extension digits when saying caller id.
; sayduration=no          ; Turn on/off the duration information before
;                              the message. [ON by default]
; saydurationm=2          ; Specify the minimum duration to say. Default
;                              is 2 minutes
; dialout=fromvm          ; Context to dial out from [option 4 from
;                              mailbox's advanced menu].
; If not specified, option 4 will not be listed and dialing out
; from within VoiceMailMain() will not be permitted.
sendvoicemail=yes         ; Allow the user to compose and send a
;                              voicemail while inside VoiceMailMain()
;                              [option 5 from mailbox's advanced menu].
; If set to 'no', option 5 will not be listed.
; searchcontexts=yes      ; Current default behavior is to search only
;                              the default context
; if one is not specified.  The older behavior was to search all
; contexts.
; This option restores the old behavior [DEFAULT=no]
; callback=fromvm         ; Context to call back from
;      if not listed, calling the sender back will not be permitted
; exitcontext=fromvm      ; Context to go to on user exit such as * or 0
;                          ;     The default is the current context.
; review=yes              ; Allow sender to review/rerecord their
;                              message before saving it [OFF by default
; operator=yes            ; Allow sender to hit 0 before/after/during
;                              leaving a voicemail to
;      reach an operator  [OFF by default]
; envelope=no             ; Turn on/off envelope playback before message
;                              playback. [ON by default]
;      This does NOT affect option 3,3 from the advanced options menu
; delete=yes              ; After notification, the voicemail is deleted
;                              from the server. [per-mailbox only]
;      This is intended for use with users who wish to receive their
;      voicemail ONLY by email. Note:  "deletevoicemail" is provided as
;      an equivalent option for Realtime configuration.
; volgain=0.0             ; Emails bearing the voicemail may arrive in
;      a volume too quiet to be heard.  This parameter allows you to
;      specify how much gain to add to the message when sending a
;      voicemail.
;      NOTE: sox must be installed for this option to work.
; nextaftercmd=yes        ; Skips to the next message after hitting 7 or
;      9 to delete/save current message.
;      [global option only at this time]
```

```
;  forcename=yes            ; Forces a new user to record their name.  A
;     new user is determined by the password being the same as the
;     mailbox number.  The default is "no".
;  forcegreetings=no        ; This is the same as forcename, except for
;     recording
;     greetings.  The default is "no".
;  hidefromdir=yes          ; Hide this mailbox from the directory
;     produced by app_directory
;     The default is "no".
;  tempgreetwarn=yes        ; Remind the user that their temporary
;     greeting is set
;  vm-password=custom_sound
;     Customize which sound file is used instead of the default
;     prompt that says: "password"
;  vm-newpassword=custom_sound
;     Customize which sound file is used instead of the default
;     prompt that says: "Please enter your new password followed by
;     the pound key."
;  vm-passchanged=custom_sound
;     Customize which sound file is used instead of the default
;     prompt that says: "Your password has been changed."
;  vm-reenterpassword=custom_sound
;     Customize which sound file is used instead of the default
;     prompt that says: "Please re-enter your password followed by
;     the pound key"
;  vm-mismatch=custom_sound
;     Customize which sound file is used instead of the default
;     prompt that says: "The passwords you entered and re-entered
;     did not match.  Please try again."
;  listen-control-forward-key=#        ; Customize the key that fast-
;     forwards message playback
;  listen-control-reverse-key=*        ; Customize the key that rewinds
;     message playback
;  listen-control-pause-key=0  ; Customize the key that pauses/unpauses
;     message playback
;  listen-control-restart-key=2        ; Customize the key that restarts
;     message playback
;  listen-control-stop-key=13456789   ; Customize the keys that
;     interrupt message playback, probably all keys not set above

; Maximum number of messages allowed in the 'Deleted' folder. If set
;     to 0
; or no then no deleted messages will be moved. If non-zero (max 9999)
;     then up
; to this number of messages will be automagically saved when they are
; 'deleted' on a FIFO basis.
```

```
; defaults to being off
; backupdeleted=100
[zonemessages]
eastern=America/New_York|'vm-received' Q 'digits/at' IMp
central=America/Chicago|'vm-received' Q 'digits/at' IMp
central24=America/Chicago|'vm-received' q 'digits/at' H N 'hours'
military=Zulu|'vm-received' q 'digits/at' H N 'hours' 'phonetic/z_p'
european=Europe/Copenhagen|'vm-received' a d b 'digits/at' HM
[default]
; Define maximum number of messages per folder for partcular context.
;maxmsg=50
1234 => 4242,Example Mailbox,root@localhost
6001 => 9999,David Duffett,dduffett@packt.com,,tz=european
6002 => 9999,Jared Smith,jsmith@packt.com,,tz=eastern
```

Changing the language of system prompts

Asterisk users know that sound files are usually kept in the /var/lib/asterisk/ sounds directory. If you look into this directory on an Asterisk 1.6.x.x installation, you will find that it does not have any sounds in it at all. It actually has three subdirectories called en, es, and fr.

As you can guess, each of these subdirectories represents a language, en (English) being the default. In fact, neither es (Spanish) nor fr (French) have any prompts on installation, but are put there to show the way things are done.

You can create as many other directories as you need, such as de, nl, it, and so on for the storage of prompts in whatever language(s) you need.

```
/var/lib/asterisk/sounds/en/
/var/lib/asterisk/sounds/es/
/var/lib/asterisk/sounds/fr/
/var/lib/asterisk/sounds/it/
/var/lib/asterisk/sounds/nl/
```

Within these directories, the system prompts must be in the file names known to Asterisk. So, the file called /var/lib/asterisk/sounds/en/hello.gsm will contain "Hello" whereas the file called /var/lib/asterisk/sounds/fr/hello.gsm will contain "Bonjour".

You may have already seen this principal at work—some prompt companies offer a standard package of Asterisk sounds in a given language or accent, which are all recorded in the standard filenames. Thus, they can be dropped into the appropriate directory, and you are ready to go.

The way you select the prompts to be used during a call is by specifying it in the concerned channel configuration file.

In `sip.conf` and `iax.conf`, you can set the language in the `[general]` section of the files:

```
[general]
language=en       ;default language
Or you can set the language for a given user:
[203]
type=friend
language=fr       ;user-specific language
```

Thus you can have a default language (as identified in the `[general]` section) and/or select a language for each specific user, and as a call comes in through that channel, the language will be set.

The same is true for DAHDI channels. You can set the default language channel-specific languages in the `[channels]` section of `chan_dahdi.conf`:

```
[channels]
language=en       ;default language
...
channels=>1-23
...
language=fr
channels=>24      ;channel-specific language
```

There you have it, a very short section, because changing the language of your prompts or using multiple language prompts in Asterisk is very easy.

Local telephony interfaces

Having set the tones, announcements, and language so that your users feel at home, the next task is to physically connect our Asterisk to the outside (telephony) world.

Fortunately, the RJ-45 is ubiquitous as far as Ethernet connectivity is concerned, so attention can be focused on traditional telephony connections.

Analog

It is a fairly safe bet that whatever equipment you choose to connect your analog lines and phones to Asterisk, whether you have gone with an ATA or a Digium card, the termination will be with an RJ-11 socket (with the middle two pins used for the pair).

This means that the lead that plugs into that RJ-11 socket becomes a mission-critical accessory, which, in a number of cases, is not shipped with your equipment.

You need to be absolutely sure that the lead you use has the right pin to pin connections as well as the right connectors at each end.

This may seem like stating the obvious, but it is included as a result of a Digium support case where a customer in the UK reported that the analog card they were using was faulty. The case got to the stage where the support team remotely accessed the machine in the UK and established that there appeared to be an unexpected voltage on one of the legs of the telephone line. When someone questioned the patch lead connecting the card to the PSTN, only then was it established that the customer had just picked up a lead that had the right ends and plugged it in, without worrying about whether the electrical connections were right. It turned out that the lead was probably for some old modem and it cross-connected some of the pins.

Of course, physical connectivity is essential, so is ensuring that the pinouts are correct. In the left part of the following figure, we can see the type of lead in question. It has an RJ-11 at one end, and a BT (British Telecom, the main telco in the UK) plug at the other. To the right is a picture of the two ends of the lead that would be required to connect to an analog telephone line in Korea. This has been included to give an idea about the diversity of connectors that are used around the world.

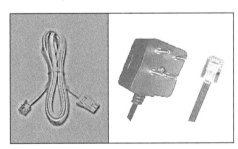

From a regulatory standpoint, it should be pointed out that only analog telephony cards and ATAs with the appropriate approvals should be connected to the telephone network in a given country.

The main learning point here is to ensure that you consider physical connectivity as part of your localization planning.

Digital

Digital telephony interfaces will be in two groups — basic rate access (otherwise known as BRI or ISDN2), and primary rate access (otherwise known as PRI, E1, T1, or ISDN, and sometimes ISDN30 in countries where 30 voice channels are used).

BRI is almost universally connected using an RJ-45 patch lead, which will look like a standard Ethernet patch, but if more closely inspected, it becomes apparent that only four wires are connected.

PRI connectivity started being implemented using two BNC connectors (one for transmit and one for receive), but is more commonly connected using RJ-45s these days.

 Note that a PRI patch lead with RJ-45s at each end will be different from a BRI patch cable, which as mentioned, is different from an Ethernet patch. So, label your cables!

In countries such as the UK, you would not expect to find signaling systems with telco presentation on co-axial connections, but many countries (for example, those found in South America) still utilize R2 signaling, which may well be presented using co-ax.

Most PRI apparatus (including telephony cards) will use RJ-45 sockets these days. Therefore, it may be possible that in some countries you are presented with a situation where the telco is giving you two BNC connectors, but your equipment wants an RJ-45 plug. This is simply remedied by using a **balun** (so named because it is connecting a balanced interface, the RJ-45 end, to an unbalanced interface, the BNC end). Here is a figure to illustrate it:

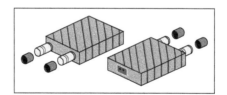

When I say "simply remedied", that is as long as you have thought of it in advance, and ordered these inexpensive, yet very useful adapters. If you have not thought of it prior to implementation and find yourself in need of a balun, be prepared to wait to get them shipped. These are not items you will find at the corner store.

Localizing caller ID signaling on Digium analog interfaces

Given the work we have already identified in this chapter to localize your Asterisk telephony solution, it will not come as a complete surprise to learn that the method used to send caller ID information over analog lines varies from country to country.

Fortunately, Asterisk is capable of recognizing many different types of caller ID signaling. This should not be a problem as long as you remember to set up the correct type for the country where the system will be connected.

All caller ID settings are made in `chan_dahdi.conf` (formerly `zapata.conf`), which is found in the `/etc/asterisk/` directory.

When you look into this file, the only lines you see concerning caller ID are:

```
usecallerid=yes
hidecallerid=no
```

Asterisk will be using its default type of caller ID signaling, which is North American. You may not even see these lines, as this is the default position.

In order to change things, you will need to add some extra lines to the file. The number of lines you add will depend on the type of signaling you want to work with, and whether you are connecting to analog phones, or lines, or both.

Shown next, is the kind of entry you would expect to see for a connection to an analog telephone line, with the entries that change the caller ID signaling highlighted.

```
context=from_outside
signalling=fxs_ks
usecallerid=yes
hidecallerid=no
cidsignalling=v23    ;BT CallerID presentation signalling method
cidstart=polarity    ;Indication of CallerID data starting
callerid=asreceived
callwaiting=no
channel => 4
```

The two highlighted lines are equipping Asterisk to recognize the caller ID as it is sent in the UK over BT (British Telecom) analog lines. The `callerid=asreceived` line is not part of the change and serves to propagate the caller ID information through Asterisk. Therefore, it will show up on connected extensions when they are called.

What we saw just now covers the case where we want to interpret incoming caller ID data. However, we also need to look at the sending of caller ID to any analog phone that we have connected to our Asterisk. If the analog phones are connected using ATAs, it is those devices that will handle the caller ID. If the phones are connected to a Digium card, then we are back in `chan_dahdi.conf`.

```
context=users
signalling=fxo_ks
callerid="David Duffett" <5001>
cidsignalling=v23   ;BT CallerID presentation signalling method
cidstart=polarity   ;Indication of CallerID data starting
sendcalleridafter=2;The number of rings before sending data
mailbox="5001"
callwaiting=yes
threewaycalling=yes
transfer=yes
channel => 1
```

In the code that we saw just now, the third highlighted line is particular to only a few caller ID signaling systems, and instructs Asterisk to send the caller ID data after (in this case) two rings. This is for systems where the data is sent between rings. It is the kind of system used in the UK.

A final note on caller ID: Do make sure that your customer actually has caller ID enabled on their telephone line before you run around worrying that Asterisk is not recognizing it. There was a guy (who may or may not be one of the authors) who spent a good amount of time analyzing why the caller ID was not showing up in Asterisk by changing settings, changing them back, and so on before asking the customers if they actually had caller ID on their line. The response was: "No, we never use it". Needless to say, it's the kind of mistake you make only once or twice.

Checklist

So, that's the end of our localization journey together. We hope you gain as much enjoyment from reading it as was gained from writing it. In closing this part of the chapter, a checklist is included so that you can be systematic in your approach to applying localizations to your Asterisk-based solutions.

Check	What to change	Where to change it
	Tones (recognized and generated) internal to Asterisk	`etc/asterisk/indications.conf`
	Tones (recognized and generated) for telephony interfaces through Digium hardware	`etc/dahdi/system.conf`
	Tones heard on SIP/IAX phones, or through ATAs	On the devices themselves
	Time and date formatting, and announcement	`etc/asterisk/voicemail.conf`
	Language of announcements	The appropriate channel, for example,. `etc/asterisk/sip.conf`
		`etc/asterisk/iax.conf`
	Physical interfaces	Be sure to get lead with the right ends and connections.
		Get baluns if required.
	Caller ID signaling (if used)	`etc/asterisk/chan_dahdi.conf`
		for Digium cards or on the ATA

Practical security

As mentioned previously, this part of the chapter must be seen as containing security measures below the headline security, which will be within your network as a whole—stuff such as firewalls.

The assumption is that your network is as secure as you want it to be. The measures identified in this section are some ideas on how to prevent abuse of your system by those coming into it on telephony channels of one sort or another.

The first thing to look at is how to stop unauthorized users hooking up to a channel on your Asterisk server. This can be done by allowing only given IP addresses or address ranges to connect with your Asterisk through, say, the SIP channel. It is easily implemented in `sip.conf` like this:

```
[my-sip-profile]
type=friend
deny=0.0.0.0/0
permit=192.168.2.200/255.255.255.255
context=internal
```

The example above would only allow a device with the IP address of `192.168.2.200` to connect to that SIP profile. Notice that we first `deny` "all", before permitting the authorized address.

To allow the 192.168.2.XXX range of addresses, the entry should look like this:

```
deny=0.0.0.0/0
permit=192.168.2.0/255.255.255.0
```

In addition to restricting the IP address of devices hooking up to your system, you also need to pay very close attention to the `[general]` section of `sip.conf` and `iax.conf`, because this section may contain a line that directs unknown callers to a specific context in the dialplan.

 A context in the dialplan is actually a mini dialplan in its own right.

In the sample configurations that come with Asterisk (which should not be used for a production system), the line looks like this:

```
context=default
```

This line may also appear in your `chan_dahdi.conf` (formerly `zapata.conf`).

This means that any of the calls that come in on those channels that do not match with one of the profiles specified lower down in the file, will be passed into the dialplan in the `[default]` context. Therefore, it is vital that you ensure—if you have a context called `[default]` in your dialplan, you don't allow callers that end up in that context to make calls that could cost you money, or waste the time of those with extensions on the system.

If you've checked your configuration and determined that you don't have a
context=default line, don't think that you're safe from this particular feature. Even if
there isn't a context=default line in the [general] section, Asterisk will still direct
any of the unknown calls to a context called [default] in your dialplan, if you have
such a context. So, it might be wise "not" to have a [default] context in your dialplan,
or to ensure that if the context is present, it just plays an "unauthorized call" message.

To make things explicitly clear, you may wish to have a [incoming_untrusted]
context to which all unknown IP calls are directed by the line context=
(in the [general] section of sip.conf or iax.conf), instead of the [default]
context. Now, a special message can be played to the calls which end up in the
[incoming_untrusted] context.

Talking of contexts, these are our first line of defense against misuse by those who
are allowed on the system. Once a call comes into a given context in the dialplan,
there is no way for it to leave that context unless you allow it with a conditional or
unconditional branch (GotoIf or Goto), or unless you have included some other
context. In this case, the call will try to find a match in the included context only if
no match is found in the first context it came into.

Knowing this enables us to use contexts to deny external callers the ability to make
outbound calls, or to build a class of access model where certain internal users
can make only locals calls. However, other, more privileged users, can make long
distance and international calls too.

We direct a call coming into Asterisk by the line context= for that particular profile
in the channel configuration file. Here is an example for a SIP phone:

```
[worker-sip-shone]
type=friend
host=dynamic
secret=thepassword
context=local
```

The highlighted line in this profile, sends any call that comes in through that profile to
the [local] context in the dialplan. The calls sent into that context might be allowed
to call local numbers, free numbers, and other extensions on the PBX. (In some time,
we'll see them as four digit numbers that always begin with a "2" in an example.)

If we look at the profile for a manager's phone on the PBX, it will look like this:

```
[manager-sip-shone]
type=friend
host=dynamic
secret=thepassword
context=ld-and-international
```

The highlighted line signifies that we can make long distance and international calls.

To see how this affects their respective abilities to make outside calls, we will need to look at the dialplan in extensions.conf:

```
[ld-and-international]
exten => _1NXXNXXXXX,1,Dial(SIP/ld-provider/${EXTEN}
exten => _011.,1,Dial(SIP/international-provide/${EXTEN})
include = > local
[local]
exten => _NXXNXXXXX,1,Dial(SIP/local-provider/${EXTEN})
exten => _NXXXXX,1,Dial(SIP/local-provider/${EXTEN})
exten => _18XX.,1,Dial(SIP/local-provider/${EXTEN})
exten => _2XXX,1.Dial(SIP/${EXTEN})
```

As you can see from the previous dialplan excerpt, the calls dropped into the [local] context can only dial local, free, and internal calls. If the dialed number does not match one of the four patterns shown, the call will be rejected as "extension not found".

Calls dropped into the [ld-and-international] context can call long distance and international destinations. Also, by virtue of the include => local part, they can make local, free, and internal calls too.

Please note that the aim here is to demonstrate the principal of restricting access, or creating classes of service, using contexts and includes. You will need to create a set of contexts that match the market you are in. For example, you may wish to limit access to mobile calls in places like the UK (where making calls to mobile phones from a landline is relatively expensive), however, this would be unnecessary in places like the US (here, the cost of making calls to mobile phones is exactly the same as that for calls made to a landline).

Another method of securing access to expensive calls is the `Authenticate()` dialplan application, which gives the user three tries to enter the specified PIN, and if unsuccessful, Asterisk hangs up the call. Successful PIN entry enables the next priority in the dialplan. Implementing this security method, for international calls made out of the US is as simple as:

```
exten => _011.,1,Authenticate(1357)
Exten => _011.,1,Dial(SIP/international-provider/${EXTEN})
```

When users dial a number starting with "011", they will be prompted for a PIN. Unless they enter "1357" within three tries, they cannot make an international call. You can use the `${}` referencing to set the PIN, rather than hardcoding it, as I have done in the previous example

There is also an application called `VMAuthenticate()`, which does the same as the `Authenticate()` application that we saw just now. However, it uses mailbox PIN codes to authenticate the users, so each of them can have their own unique PIN (as long as they have a voicemail box).

This security measure is great for hotel house phones or phones on the reception desks of businesses. Here, it would be good to have the convenience of the staff being able to use them for say, international calls, but you don't want anyone who wanders in off the street being able to do the same.

There are, of course, a good number of more elegant or sophisticated ways to implement user authentication within Asterisk (perhaps including the use of the AstDB database). However, here the aim has been to highlight a couple of ready-made applications within Asterisk.

Out of hours

The last practical security measure we are going to look at is out of hours calling. Let's assume you, or your customer, want to implement a mechanism to secure office phones, which can currently make unlimited calls at any time of the day or night. To achieve this, you will require authentication for all calls placed between 6.00 pm and 7.30 am during the working week, and for all weekend calls. This prevents the friendly cleaning staff, or anyone else, from making expensive calls during this timeframe.

The next example, would allow a given phone to make outside calls (identified by the prefix "9", which we strip away from the front of the number before calling it) during the working day, but outside of those hours, a PIN (of 1357) will be required to give access to outside calls:

```
exten => _9.,1,GotoIfTime(07:30-18:00,mon-fri,*,*?allowed)
exten => _9.,n,Authenticate(1357)
exten => _9.,n(allowed),Dial(SIP/my-sip-trunk/${EXTEN:1})
```

Therefore, when a call is placed (with a leading "9") between the hours of 07:30 and 18:00, Monday to Friday, we jump to the priority label "allowed", which makes the call.

Outside of these hours, the call just drops down to the next priority, which prompts the user to enter a PIN before allowing the call to be made. Two extra lines in our dialplan could easily save the company a lot of money!

Summary

The most important thing to do concerning the localization of the telephony systems you implement, is to walk in your customer's (and their customers') shoes. This means that you go through every contact and scenario that they would go through with the system (making a call, retrieving a voicemail, and so on) to ensure that everything sounds and appears **normal** to them.

A point that is made well in the other chapters of this book is that customers will have certain expectations when they pick up a phone. If that phone is on your system, then it's your job to meet those expectations such as familiar dial tone, local language/accent announcements, and so on.

If you're reading this book, you will be the type of person who wants to exceed expectations, and that's great, but these are not the places to do it, people don't want surprises on stuff like this.

The great news is that with the unparalleled (almost infinite) flexibility of Asterisk, you're sure to be able to impress your customers with the added functionality, the seamless fit in to their business, and the overall elegance of the solutions you implement for them.

Having covered a good number of areas, such as tones, physical interfaces, and announcements, it is reassuring to know that, although Asterisk was born in the USA, with a few minor changes to some configuration files and the right connectors in place, "you're all set" (as they say in the US) for successful international deployments.

On security, however, Asterisk must be handled at a network level. There is also scope to implement some security measures within Asterisk—both regarding who is allowed to connect to Asterisk, and which calls they are allowed to make once an authorized user is connected.

With these measures being so simple to implement, please put them in as a precaution at the creation of your solution. Don't let them be a reaction to an "issue" that was allowed to occur.

9
Interfacing with Traditional Analog and Digital Telephony

As a well-known book tells us, Asterisk is the future of telephony. In addition to being an open-source project that uses open standards to enable any number of IP telephony applications, one of the things that ensures Asterisk's future is its ability to connect with the telephony interfaces of the past, which are still very much in use today.

Before we discuss the options that exist to connect Asterisk to the traditional telephony interfaces, let's take a quick (overview-level) look at each of those which you might come across.

If you already know the basics of how analog and digital telephony work, feel free to skip ahead to the *Choices, choices* section of this chapter.

Analog

This is the simplest interface you will encounter. It is ubiquitous as there are not many places you will go to that do not support analog telephony. Indeed, you probably have at least one of these connections in your home or office. It is sometimes known as **Plain Old Telephone Service (POTS)**, denoting its simplicity. I once heard **Voice over Internet Protocol (VoIP)** described as Pretty Amazing New Stuff, you may smile at the corresponding acronym.

With this interface, we connect a standard analog telephone (cheap, easy to obtain and use) to a standard line, which generally arrives on a twisted pair. The capability is easy to understand. You can make one phone call and then the line is busy. If you want to run two calls at once, you need to get another line.

The phone is known as a **Foreign eXchange Subscriber** (**FXS**) device and in the absence of a PBX, it plugs into the phone line from your friendly telco. In this nomenclature, the telco local exchange that you connect to is known as a **Foreign eXchange Office** (**FXO**).

This diagram should make things clear:

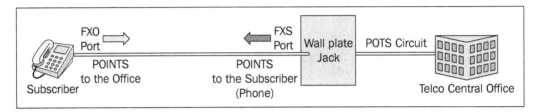

Analog lines and phones are commonly used in both domestic and business situations. These lines are not what one would call feature-rich, but they can support a number of services like call waiting, three-way calling, and caller ID presentation (caller ID transmission methods vary around the world—Asterisk supports many formats, for more details, see Chapter 8 on localization).

There are two main methods used by analog lines to signal to the exchange that the phone has gone off-hook, and therefore, that a dial tone is required. One of these methods is called **ground start** (or Earth calling), which simply grounds one leg of the twisted pair. The other is called **loop start**, which simply loops the pair. Loop start signaling is also used by an enhanced system which boasts far end disconnect supervision and this system is known (in Asterisk, at least) as **Kewl Start**.

It is important that you know which system is in use on the lines you plan to connect to Asterisk.

Once the dial tone is on the line, the user can dial the desired number. In the old days (think rotary phones), the number was transmitted to the exchange by a method called **loop disconnect** (or **pulse dialing**)—the pair of wires is temporarily shorted together at a speed of 10 times per second to each digit. One pulse represented the number "1" all the way through to nine pulses for "9", and ten pulses for "0".

These days most analog lines use DTMF for dialing. Here, a unique combination of two pure tones is used to represent each digit, as shown in the following diagram:

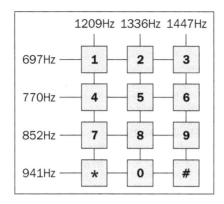

One small but important note is that in the usual run of things, there is no way for Asterisk to determine which number was dialed when an analog line rings, as there is for IP telephony, BRI, or PRI lines. This means that when a call is received from an analog line into Asterisk, there is no extension number that we could match it against. Hence, we use the "s" (or start) extension, for example:

```
[from-analog-line]
exten => s,1,Background(welcome-menu)
```

Businesses with a need for a number of lines can either use a collection of analog lines, or they may choose to go for a digital connection (described next) which can provide a number of concurrent calls together, along with some additional services.

Digital

This kind of interface provides a number of channels for conveying digitized voice information (these are known as bearers). The voice is sampled 8,000 times a second, and each time it is sampled, an 8-bit representation of the amplitude of the voice is made. With eight bits of information being generated 8,000 times a second, we get 64,000 bps (bits per second), and this process is known as **PCM (Pulse Code Modulation)**. Thus the standard way of digitizing the voice in telephony is called PCM64.

If you are wondering why the voice is sampled at the rate of 8,000 times per second, it was one Harry Nyquist who devised the sampling theorem that took his name. It states that, in order to faithfully recreate an analog waveform at the far end of a digital transmission, it must be sampled at a rate which is at least twice as fast as the highest frequency being sampled.

To put some practical numbers on this, most analog telephone lines are guaranteed to transmit signals up to 3,400 Hz (or 3.4 kHz), so to faithfully recreate these signals, they need to be sampled at least twice as fast as 3,400 times per second, which is 6,800 times per second. We then add some magical fiddle factor, which no one seems to fully understand to get the deemed standard of 8,000 times per second.

There are two ways of arriving at the 8-bit samples that are in use at the time of writing of this book. One is called a-law, and is used in Europe and many other parts of the world. The other is called μ-law, and is dominantly used in the US, Canada, and Japan. Both are variants of the G.711 codec.

Once we have the voice in the digital form, we have the opportunity to multiplex a number of "digital voices" onto a single transmission line, and this is exactly what we do by a process known as **time-division multiplexing (TDM)**. A simple way of understanding this is to think of carriages on a train. We put a single one of those voice samples (the 8-bit sample described just now) into each carriage, and, provided the far end knows which carriage is which, the samples can be taken out of the carriages and reconstructed into the individual telephone conversations involved.

There are two main types of digital telephony interface that you will come across—one is called the **Basic Rate Interface** (or **BRI**) and the other is called the **Primary Rate Interface** (or **PRI**). Both are **ISDN (Integrated Services Digital Network**) links.

ISDN BRI (Basic Rate Interface)

BRI provides a digital connection with two 64 kbps bearer channels and one 16 kbps bearer channel, so the data rate of this type of a connection is 14,400 kbps.

BRI arrives at the customer premises from the telco on two wires (this is known as the U bus), and the network termination and test equipment supplied by the telco (the box that you plug your equipment into) turns it into a **four-wire system**—a transmit pair and a receive pair, usually known as the S/T bus. In addition to supporting two concurrent calls, BRI also provides some other services, including conveying the number that was dialed by the caller, the caller's number (caller ID). BRI also has a facility to host a number of individual telephone numbers, which will arrive on that single BRI connection. The numbers may be known as **direct inward dialing (DID)**, **direct dial-in (DDI)**, or **multi-subscriber numbering (MSN)**, depending on where you are in the world and the way the scheme is implemented.

The equipment connected to the BRI is able to see which number was dialed by the caller, and so it is possible to route the call based on the dialed number.

BRI is very popular in a number of European countries like Germany and the Netherlands. Its popularity is limited in some countries, where the service providers do not facilitate broadband connections over it, as is the case with ADSL over analog lines. It is not at all popular in the US. Here is a diagram of a BRI.

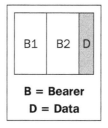

If you would like more detailed information on the BRI interface, there is a great resource at `http://ckp.made-it.com/isdn.html`.

ISDN PRI (Primary Rate Interface)

A PRI connection will provide a large number of voice channels, namely, 23 in the US and Japan where the link is called a T1, and 30 in Europe and many other regions where the link is called an E1. Both types have a data channel, which is 64 kbps as in the case of voice channels. When all the information regarding all of the calls on the 23 or 30 voice channels is exclusively conveyed in the data channel, it is what we call a **Common Channel Signaling (CCS)** protocol.

 It is possible to get 24 channels on a T1 using what is called a **robbed bit protocol**, sometimes known as **T1RB,** where the data that would have been transmitted in the data channel is instead transmitted over a number of frames, using the least significant bit of the samples in the voice channels. Thus, what was the data channel can be used for voice too.

Older systems will transmit the call set-up, the tear-down messaging, and so on as tones within the individual voice channels. These systems are said to use a **Channel Associated Signaling (CAS)** protocol.

You will remember the brief mention of G.711 a-law and μ-law audio encoding. Well, a-law is traditionally used on an E1 link, and μ-law is traditionally used on a T1 link.

Shown next are pictures of a T1 (both standard and robbed bit can be seen) and an E1 link, together with some example protocols that may be used on them.

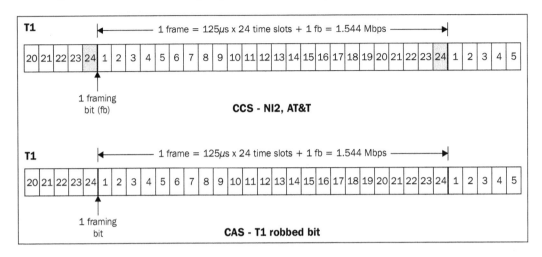

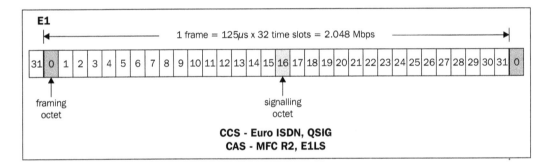

You will see from the diagrams shown just now that each primary interface runs at a speed dictated by the number of channels.

With both BRI and PRI trunks, there are a number of aspects of the connection, which will need to match up with parameters set on the equipment to which they will connect. First is the fact that both systems work on the principal two complementary ends connecting to the circuit—in the case of PRI, the ends are variously known as Master and Slave, Network and User, Ax and By, and so on depending on what flavor of protocol is in use. In the case of BRI, the ends are usually known as NT for the network end, and **Terminating Equipment** (TE) for the user end.

When using interface cards that work within the DAHDI framework, Asterisk uses the terms `net` (for network end) and `cpe` (or Customer Premises Equipment, for the user end) in combination with the type of interface, to specify the type of signaling in `/etc/asterisk/chan_dahdi.conf`. For example, one of the following would be used to describe a given interface:

```
signalling=pri_cpe
signalling=pri_net
signalling=bri_cpe
signalling=bri_net
```

Without complimentary ends being configured on the circuit (that is, if you are connecting to the telco, your equipment must be set up to be the user end), your link will not work.

Another important thing to consider is timing — since both these are time-division multiplexed systems, it is crucial to ensure that timing is set up correctly.

One end must give the synchronization and the other end must receive the synchronization. The organization of this (synchronization) is usually implied when you set whether you are the network end or the user end. If you are connecting to a telco interface, then it is normal for you to be considered the "user", and therefore, you will receive the synchronization.

You would choose to generate the synchronization locally (and supply it to the other end), if you were connecting your solution to the telco interface of some other equipment (which means your solution was "pretending to be the telco").

If both ends are trying to give synchronization, or if both ends are trying to receive synchronization, your link will not work.

With these two parameters correctly set, the physical link should be fine, and your next job is to ensure that the signaling system, framing, and encoding match with the other end.

If you choose to use an external gateway with Asterisk to connect to PRI or BRI interfaces, all of the parameters stated previously will need to be set through the user interface of that equipment.

If you're using interface cards that work through the DAHDI framework, some of these parameters will be set in `/etc/dahdi/system.conf` and the rest will be set in `etc/asterisk/chan_dahdi.conf`. Further details are in the *Installing a Digium card* section of this chapter.

Choices, choices

The fundamental choice that we have to make when looking to connect any of the above traditional telephony interfaces, is whether we are going to use an external adaptor (best known as a gateway) or an internal card.

There are pros and cons with both options, so this table has been included to compare them:

Considerations	External adaptor	Internal card
Density	A good number of sizing options available from single or multiple FXSs to a number of BRIs and PRIs.	A good number of sizing options available from single or multiple FXSs to a number of BRIs and PRIs.
Power supply	Almost always a separate module like a plug-top adaptor.	It usually takes power from the host platform from the bus and (in the case of FXS ports which need to supply line and ringing voltages to the phone) the other voltages available. External PSU will be used for high density FXS implementations.
Form factor	External, so no issue.	The card must be perfectly compatible with the interface offered by the platform. For example, PCI, PCI-X, PCIe, or cPCI in a PC, or maybe a bespoke interface in an embedded platform.
Points of failure	PSU, the adaptor.	The card and compatibility with the host platform (bus voltage, IRQ).
Neatness	Lots of leads, for example, an ATA (which might convert a single FXS device to SIP) will require the power lead, the lead to the device, and the Ethernet lead to the Asterisk host PC.	It's very neat. The device or connection plugs straight into the socket on the card (high density FXS/FXO solutions may use an external breakout box).
Convenience	It's very convenient, no tools required—just hook up the cables.	Need to power down and open the PC or embedded platform to install the card.

Considerations	External adaptor	Internal card
Ease of installation	Very easy — the adaptor (usually) turns the traditional connection into SIP or IAX, which is well supported by Asterisk, and easily configured.	Will require a driver to make the card work under the operating system, and also some editing of the appropriate `.conf` file to link the channels to Asterisk.
Portability (in terms of moving the device to another host)	Very easy.	Must ensure the new host is totally compatible with the old. This presents an issue — for example, if you have been using a PCI card, you must find a new host with a compatible PCI slot.

Putting some thought into the way to go for a given project is time well spent. There are some implementations where cards will be the best option and other scenarios where an external adaptor will be the way to go.

Using external adaptors

If you choose to go with external adaptors, you will need to configure both the adaptor and Asterisk, to hook up with the adaptor (usually, as already mentioned, in `sip.conf` or `iax.conf`).

Here is a diagram to show an FXS adaptor, known as an **ATA** (**Analog Telephony Adaptor**) within the context of a network diagram, so that you can see how it fits in:

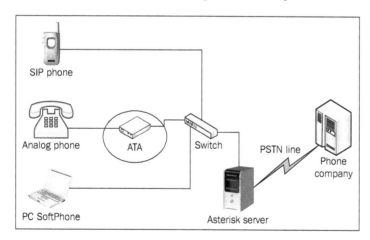

In the example shown, you can see the ATA has the FXS connection on one side and an Ethernet LAN connection (going to the switch) on the other side.

Of course, these adaptors are also available for FXO, BRI, and PRI connections in many different levels of density. See the end of this section for some names of companies that manufacture external adaptors that can be used with Asterisk.

External adaptors are virtually always configured through a web interface, which means we need to know the default IP address of the adaptor to get into it in the first place and access the configuration menu. We will also need to reset the static IP address to one which fits with our network, or set it to be a DHCP client.

 Hint: If setting the device to a static IP address, it's a great idea to get your labeling machine out and print out a label with that IP address and the username and password for the web interface, and stick it on the device BEFORE you do anything else. This avoids having to guess the IP address at a later date, along with the frustration of forgotten credentials after you have eventually found the IP address. Of course, this does assume your device is located in a physically secure area.

Some (which are not specifically designed to work with Asterisk) may have a WAN connection so that they can act as the Internet router, this allows them to impose a **QoS (Quality of Service)** scheme in favor of the IP telephony traffic.

There will be a number of settings, or pages of settings, of which the majority can probably be left untouched. It is not proposed to go through all the conceivable options, but there are several very important settings that you will need to configure.

For the traditional connection, whether FXS, FXO, BRI, or PRI, you will need to set the device to match what you are hooking it up with. Stuff like line impedances and caller ID presentation for analog connections will usually be set up in a specific section.

The other settings that are a must are the IP telephony settings, so that you can hook the unit up with the appropriate SIP or IAX profile on your Asterisk.

There follows a (far from comprehensive) table listing companies along with their web sites, which make adaptors of one sort or another:

Company	Web site
VegaStream (gateways supporting FXS, FXO, BRI, or PRI)	`http://www.vegastream.com`
Redfone (TDMoE bridges/PRI gateways)	`http://www.red-fone.com`
Patton (very comprehensive range of gateways)	`http://www.patton.com`
Digium (creator, sponsor, and primary developer of Asterisk)	`http://www.digium.com`
Xorcom (a very flexible range of USB2.0 connected gateways)	`http://www.xorcom.com`

Two of the companies get a special mention:

- Digium, the primary sponsors of Asterisk, manufacture an excellent range of interface telephony cards. Each time a Digium card is purchased, revenue goes back into the development of open source Asterisk.

- Xorcom have created a very interesting range of adaptors for FXS, FXO, BRI, and PRI connections that connect to Asterisk using USB2.0. The channels show up as DAHDI channels in Asterisk.

Using cards

Some people are put off cards because they fear the installation (and possible troubleshooting) requirements within Linux. There is no need for this fear. Once you understand the way things hang together, the installation, configuration, and use of these channels within Asterisk is a fairly simple matter.

Another aspect is the fact that gateways lend themselves to redundant solutions so readily. Cards can be used in redundant systems, but only with the introduction of some kind of external switch-over apparatus.

There are lots of different companies who make cards that work with Asterisk, ranging from those who have been in the industry for years prior to the arrival of Asterisk (like Pika Technologies from Ottawa), to Digium (the natural fit for Asterisk) and a host of similar offerings from various locations in the Far East.

The Digium range of cards will cover all your needs, from one FXS or FXO connection all the way up to quad PRI (120 channels if used as E1s), and they are available in the PCI (3.3V or 5V) or PCIe form factors. There is also a quad BRI (or ISDN2) card.

Installing a Digium card

The following section may seem simplistic for a book aimed at those with Asterisk experience. It is included for two reasons:

1. In the author's experience, many people who use Asterisk and are familiar with constructing dialplans and the like, are less familiar with the way `/etc/dahdi/system.conf` (formerly `/etc/zaptel.conf`) and `/etc/asterisk/chan_dahdi.conf` (formerly `/etc/asterisk/zapata.conf`) work together to enable the hardware within the Linux-based PC and support the channels within Asterisk. This can lead to difficulty when troubleshooting, so one of the aims of the book is to acquaint the reader with knowledge that will serve to accelerate the troubleshooting process.

2. It provides an opportunity to explain the new DAHDI framework, and to demonstrate the way the Linux drivers are now in a separate package to the tools needed to configure and monitor the operation of the cards.

Bearing these two points in mind, let's have a look at how to install a Digium interface card.

When any card is installed in a PC, a driver will be needed to enable the card to work under the operating system in use (you will have seen graphics cards that come with various drivers for Windows flavors, Linux, MAC OS X, and so on).

The first thing to understand is that this driver has nothing to do with Asterisk, as its job is purely to get the card working within the operating system—in our case, a Linux distribution of one sort or another.

Although Digium cards have been successfully installed under many operating systems, Digium only support their use under Linux distributions and will, in fact, give free support to anyone who buys a Digium card, to the point of getting the card functioning correctly in the platform. The drivers for Digium cards are now part of a package called **DAHDI** (**Digium Asterisk Hardware Device Interface**). In fact, we think that this is the very first book published that covers DAHDI.

Until recently, the package was called Zaptel, but the name has been changed due to copyright issues. DAHDI is the only option to use with Asterisk 1.6, and it can also be used with Asterisk 1.4, although 1.4 also supports the legacy Zaptel framework. At the same time as changing the name, Digium have cleverly separated the two parts that form the package. One is the actual Linux kernel driver, and the other is a set of tools and utilities that allow you to configure and test the cards.

DAHDI Linux (at version 2.1.x at the time of writing) is the package that contains the drivers. DAHDI tools (also at version 2.1.x) is, as its name suggests, the package that contains the tools we will use to configure and test the card(s). In addition to DAHDI, we will also need **Libpri**, a library covering the PRI specification, if we are going to install any T1/E1 cards. All of these packages are freely available at `www.asterisk.org`.

When you become familiar with the installation process for these two DAHDI packages and with the way they work together, you may wish to install both at the same time. This can be achieved by using an integrated package called `dahdi-linux-complete-2.1.x.x+2.1.x.x.tar.gz`, which is available at: `http://downloads.digium.com/pub/telephony/dahdi-linux-complete/`

For those needing detailed information covering the upgrade from Zaptel to DAHDI, there is a file called `Zaptel-to-DAHDI.txt` inside the `Asterisk-1.6.x.x` install directory created when the Asterisk tarball is untarred.

Card installation—physical

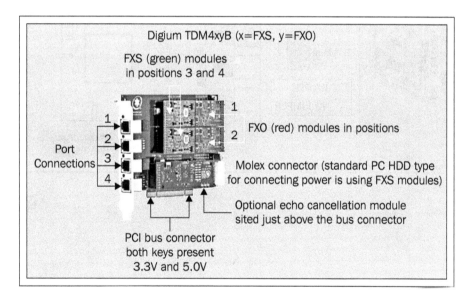

Here is the first example card we will cover, the Digium TDM422B (FXS port "2" indicates that the card has two FXS connections, the red modules FXO "2" indicates that it also has two FXO connections). This card can accommodate up to four modules, in any combination of FXS and FXO.

When installing a card, remember to power down the PC first and to take the necessary anti-static precautions. After gaining access to the appropriate slots, insert the card and, if you are using FXS modules, be sure to attach an HDD power cable from the PC to the Molex connector on the Digium card. (If you forget this, your FXS ports will not work and your card may even deny that it has any FXS modules on it.)

When installing the B410P BRI card, do remember to set the NT/TE (which changes the physical pinouts of the port to be either the Network end or Terminating Equipment end), jumpers, and the termination switches (used to add a terminating resistance to the circuit, required for long cable runs) to the desired positions.

You will set a given port to NT if you want to connect equipment (like phones) to it, and you would set the port to TE if you want to connect it to the telephone network. In other words, it has to be set to the opposite of what you will be connecting to it.

The next diagram shows Digium B410P:

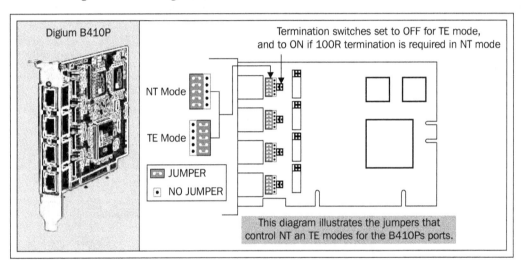

If it is a TE122P PRI card (see the next screenshot) that you want to install, the only physical thing that can be changed on the card is the jumper that sets the interface to be either an E1 (Europe and a lot of places in the world) or a T1 (North America and Japan).

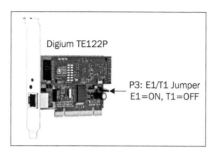

Installing the DAHDI software

Now reassemble and power up the PC, and if you have not already installed DAHDI, get to the Linux command line to do the following:

[If you have already installed DAHDI, skip down to the section called *Configuring the DAHDI files to enable the card*.]

Type the command lspci (which means list the PCI cards) and you should see the Digium card in the list, the following screenshot shows the list:

```
[root@localhost ~]# lspci
00:00.0 Host bridge: Intel Corporation 82865G/PE/P DRAM Controller/Host-Hub Interface (rev 02)
00:02.0 VGA compatible controller: Intel Corporation 82865G Integrated Graphics Controller (rev 02)
00:1d.0 USB Controller: Intel Corporation 82801EB/ER (ICH5/ICH5R) USB UHCI Controller #1 (rev 02)
00:1d.1 USB Controller: Intel Corporation 82801EB/ER (ICH5/ICH5R) USB UHCI Controller #2 (rev 02)
00:1d.2 USB Controller: Intel Corporation 82801EB/ER (ICH5/ICH5R) USB UHCI Controller #3 (rev 02)
00:1d.7 USB Controller: Intel Corporation 82801EB/ER (ICH5/ICH5R) USB2 EHCI Controller (rev 02)
00:1e.0 PCI bridge: Intel Corporation 82801 PCI Bridge (rev c2)
00:1f.0 ISA bridge: Intel Corporation 82801EB/ER (ICH5/ICH5R) LPC Interface Bridge (rev 02)
00:1f.1 IDE interface: Intel Corporation 82801EB/ER (ICH5/ICH5R) IDE Controller (rev 02)
00:1f.2 IDE interface: Intel Corporation 82801EB (ICH5) SATA Controller (rev 02)
00:1f.5 Multimedia audio controller: Intel Corporation 82801EB/ER (ICH5/ICH5R) AC'97 Audio Controller (rev 02)
05:02.0 Ethernet controller: Broadcom Corporation NetXtreme BCM5782 Gigabit Ethernet (rev 03)
05:09.0 Ethernet controller: Digium, Inc. Unknown device 8005 (rev 11)
[root@localhost ~]#
```

If the card appears in the list, great! If not, check the seating of the card in the slot.

Next, download the Libpri, DAHDI Linux, and DAHDI tools packages into the /usr/src directory, if you have not already done so.

You may ask, "Why install Libpri when this is an analog card?", great question; we tend to install Libpri along with the DAHDI software in case we do install a PRI (T1/E1) card in the future.

All these files are available as "tarballs" from www.asterisk.org.

To extract them, simply type the following at the command line (substituting the version number for the x.x):

```
# tar -zxvf libpri-1.x.x.tar.gz
# tar -zxvf dahdi-linux-2.x.x.tar.gz
# tar -zxvf dahdi-tools-2.x.x.tar.gz
```

This process will extract the contents of these gzipped tarballs into the new subdirectories of /usr/src named after each file. So, now to build and install the software type:

```
# cd libpri-1.x.x
```

Once inside the directory, type the following command:

```
# make clean (cleans up any unnecessary files)
# make install (installs the software on your system)
```

It is all that is needed here, so now we can continue further by typing (make sure your machine is connected to the Internet, as some files will be downloaded as part of this next process):

```
# cd ../dahdi-linux-2.x.x
# make (builds the source files)
# make install (installs the software on your system)
```

This process will terminate when you see the following on your screen:

```
#################################################
###
### DAHDI installed successfully.
### If you have not done so before, install the package
### dahdi-tools.
###
#################################################
[root@localhost dahdi-linux-2.1.0.3]#
```

And the dahdi-linux kernel modules are now built and installed on your system, so you can now type:

```
# cd../dahdi-tools-2.x.x
# ./configure (this checks that all necessary dependencies are installed)
# make menuselect (this invokes a simple menu [below] where you can
choose the tools to install)
```

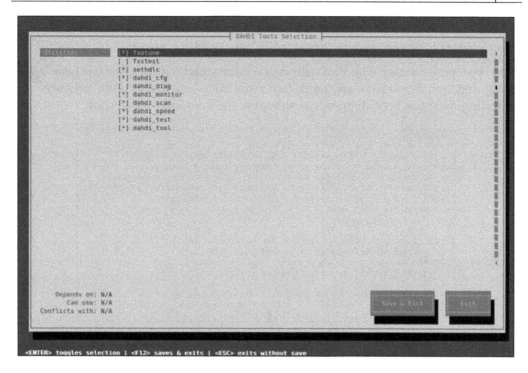

Use the right arrow key to move from the word **Utilities** to the list and then the up/ down arrows to traverse the list. *Enter* toggles the selection on (indicated with a "*") or off.

Use *F12* to save and exit.

You can use the default selections here, so you do not have to change anything unless you really need to.

Once you have exited this screen, back at the Linux command line you type:

```
# make (builds the source files)
# make install (installs the software on your system)
```

You will then be prompted (see the next screenshot) to:

```
##################################################
###
### DAHDI tools installed successfully.
### If you have not done so before, install init scripts with:
###
###    make config
###
##################################################
[root@localhost dahdi-tools-2.1.0.2]#
```

Continue further by typing:

make config (which starts DAHDI when the machine boots)

When you perform this step, you will see a confirmation that DAHDI has been configured, and if you have any hardware installed, you will also see a message indicating what has been detected, as shown in the following screenshot:

```
[root@localhost dahdi-tools-2.1.0.2]# make config
install -D dahdi.init /etc/rc.d/init.d/dahdi
/usr/bin/install -c -D -m 644 init.conf.sample /etc/dahdi/init.conf
/usr/bin/install -c -D -m 644 modules.sample /etc/dahdi/modules
/usr/bin/install -c -D -m 644 modprobe.conf.sample /etc/modprobe.d/dahdi
/usr/bin/install -c -D -m 644 blacklist.sample /etc/modprobe.d/dahdi.blacklist
install -D ifup-hdlc /etc/sysconfig/network-scripts/ifup-hdlc
/sbin/chkconfig --add dahdi
DAHDI has been configured.

If you have any DAHDI hardware it is now recommended you
edit /etc/dahdi/modules in order to load support for only
the DAHDI hardware installed in this system.  By default
support for all DAHDI hardware is loaded at DAHDI start.

I think that the DAHDI hardware you have on your system is:
pci:0000:05:09.0    wcb4xxp-    d161:b410 Digium Wildcard B410P
[root@localhost dahdi-tools-2.1.0.2]# █
```

 Asterisk uses DAHDI-enabled hardware to provide a timing source for applications that mix audio such as the MeetMe() and MusicOnHold(). Even if no cards have been installed, DAHDI will still supply timing using the dahdi_dummy driver, so this will be automatically installed in the absence of any cards.

You can now either reboot the machine or (as long as you did the make config) type:

service dahdi start

This will invoke the drivers for your Digium cards. At this point, you should see the drivers start.

```
[root@localhost dahdi-tools-2.1.0.2]# service dahdi start
Loading DAHDI hardware modules:
  wct4xxp:                                        [  OK  ]
  wcte12xp:                                       [  OK  ]
  wct1xxp:                                        [  OK  ]
  wcte11xp:                                       [  OK  ]
  wctdm24xxp:                                      [  OK  ]
  wcfxo:                                          [  OK  ]
  wctdm:                                          [  OK  ]
  wcb4xxp:                                         [  OK  ]
  wctc4xxp:                                        [  OK  ]
  xpp_usb:                                        [  OK  ]

Running dahdi_cfg:                                 [  OK  ]
[root@localhost dahdi-tools-2.1.0.2]# █
```

The above list shows which kernel driver modules have been loaded. Each module serves one or more cards in the Digium portfolio. By default, all driver modules load, but for production systems it is recommended that /etc/dahdi/modules is edited so that only the driver modules required for the actual DAHDI hardware installed are loaded.

 If you had installed Asterisk prior to going to the above steps to install the DAHDI packages, you will now need to re-install Asterisk in order for Asterisk to recognize that DAHDI is now present and install the files that will have been skipped due to the previous lack of a timing source.

Configuring the DAHDI files to enable the card

To enable the TDM422 analog card we just installed, we need to edit /etc/dahdi/system.conf — this file configures the hardware for correct operation under the Linux OS, and is NOT connected with Asterisk. Thus it resides in its own directory structure.

Inside this file, we need to specify the type of interface for each channel number. In the case of the TDM422 card, channels 1 and 2 are FXS ports and so, as mentioned previously, we need to load FXO firmware to them, because in order to work with the connected telephone, they must appear like a telephone exchange. By the same token, the channels 3 and 4, which are FXO ports, will use FXS firmware as they must appear like telephones in order to interact correctly with the local exchange.

```
fxo=1-2
fxs=3-4
loadzone=us
defaultzone=us
```

In system.conf, we can also configure the tones that will be played to and recognized from all the hardware channels (with the "zone" lines included as seen in the previous code), see the chapter on localization for details on this.

Now we are ready to use dahdi_cfg to apply the newly edited /etc/dahdi/system.conf to the hardware with the command:

#dahdi_cfg -vvv

We apply the three v's to give increased verbosity during the operation. This means that we get more information, so if things don't work out as we planned, we will have more diagnostic information.

If all is well, you should see something like this:

```
[root@localhost ~]# dahdi_cfg -vvvvv
DAHDI Tools Version - 2.1.0.2

DAHDI Version: 2.1.0.3
Echo Canceller(s):
Configuration
======================


Channel map:

Channel 01: FXO Loopstart (Default) (Echo Canceler: mg2) (Slaves: 01)
Channel 04: FXS Kewlstart (Default) (Echo Canceler: mg2) (Slaves: 04)

2 channels to configure.

Changing signalling on channel 1 from FXO Kewlstart to FXO Loopstart
Setting echocan for channel 1 to mg2
Setting echocan for channel 4 to mg2
[root@localhost ~]#
```

Our next job is to test the installed hardware for successful operation within Linux, before we even think about Asterisk.

For this very purpose, the easy-to-use utility, dahdi-tool, was created. To run it, just type `dahdi-tool` at the command line:

 # dahdi-tool

You should now see a simple screen appear that will show you the installed hardware and its status.

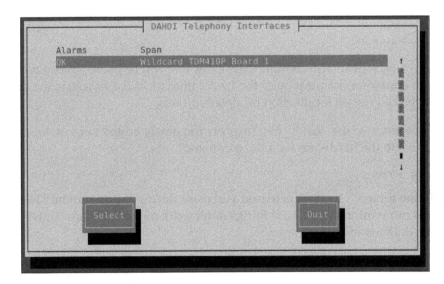

This simple tool is used only with the keyboard, so you can just use the up/down arrows and the *Tab* key to get to the thing you want and hit the *Enter* key.

When you have highlighted the analog card (which will be automatically highlighted if it is the only card in the system), tab to **Select** and hit Enter. Dahdi-tool will display some useful information about the card. The **Total/Conf/Act** line is of particular interest to us:

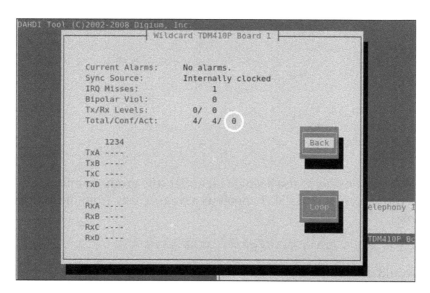

Test by lifting the handset of one of the connected phones to ensure the "active" number increments by one.

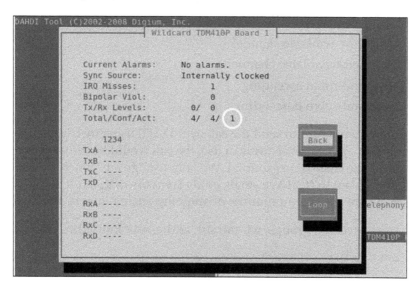

So, now we have proved the card is functioning correctly within Linux, we can now edit /etc/asterisk/chan_dahdi.conf to enable Asterisk to use those channels.

All we need to do to enable those channels is add the following lines inside the [channels] section:

```
[channels]
. . .

. . .
group=1
signalling=fxo_ks
context=users
channels=>1-2
group=2
signalling=fxs_ks
context=from_outside
callerid=asreceived
channels=>3-4
```

The group= lines allow us to use a single identifier (the group number) to address the specified selection of DAHDI channels as a single entity from the dialplan. For example:

```
exten => _9.,1,Dial(DAHDI/g2/${EXTEN:1})
```

This line would direct all calls dialed with a leading 9 (which would be removed before the number was dialed) out over the DAHDI channels identified a group 2, which in our case are channels 3-4.

Note that the g does not mean "group", it declares the strategy to use on the group. The following strategies are recognized:

- g: Use lowest available channel
- G: Use highest available channel
- r: Use round-robin ascending
- R: Use round-robin descending

So, the line above would try to send the call over DAHDI channel 3 (the lowest available in group 2). If channel 3 was in use, the call would be directed over channel 4 (the next available channel in group 2). We only use "group" when we have multiple channels that we want to treat as a single entity from the dialplan. This is very useful for situations where you have a number of outgoing analog lines or PRI channels.

Even if we have specified groups, we can still address an individual channel:

```
exten => 6001,1,Dial(DAHDI/1,20)
```

What about the other cards?

The procedure for editing `/etc/dahdi/system.conf` and `/etc/asterisk/chan_dahdi.conf` is the same when enabling the BRI (B410P) and PRI (TE122P) cards, although the actual entries differ.

In common with the analog example just seen, we use the entries in `/etc/dahdi/system.conf` to specify the attributes of each interface and which channel numbers are going to be used (because we are now configuring a digital interface, we will need to identify the data channel in addition to those that will be used for voice) and the type of echo cancellation to be used. It is in `/etc/asterisk/chan_dahdi.conf` that we specify the kind of signaling system in use, and which end of the circuit each physical interface is going to be.

In `/etc/dahdi/system.conf` the following formula is used:

```
span=<span num>,<timing source>,<line build
      out(LBO)>,<framing>,<coding>[,yellow]
```

The `<span num>` (span number) parameter is simply the numeric identifier for the individual port, starting with "1" for the first port on the first card. If you have two 4 port cards, the first port on the second card would be "5" and so on.

The following text is taken from `/etc/dahdi/system.conf` and is included as it explains the remaining span parameters so well:

###

All T1/E1/BRI spans generate a clock signal on their transmit side. The `<timing source>` parameter determines whether the clock signal from the far end of the T1/E1/BRI is used as the master source of clock timing. If it is, our own clock will synchronize to it. T1/E1/BRI connected directly or indirectly to a PSTN provider (telco) should generally be the first choice to sync to. The PSTN will never be a slave to you. You must be a slave to it.

Choose 1 to make the equipment at the far end of the E1/T1/BRI link the preferred source of the master clock. Choose 2 to make it the second choice for the master clock, if the first choice port fails (the far end dies, a cable breaks, or whatever). Choose 3 to make a port the third choice, and so on. If you have, say, 2 ports connected to the PSTN, mark those as 1 and 2. The number used for each port should be different.

If you choose 0, the port will never be used as a source of timing. This is appropriate when you know the far end should always be a slave to you. If the port is connected to a channel bank, for example, you should always be its master. Likewise, BRI TE ports should always be configured as a slave. Any number of ports can be marked as 0.

Incorrect timing sync may cause clicks/noise in the audio, poor quality or failed faxes, unreliable modem operation, and is a general all-round bad thing.

The line build-out (or LBO) is an integer, from the following table:

```
0: 0 db (CSU) / 0-133 feet (DSX-1)

1: 133-266 feet (DSX-1)

2: 266-399 feet (DSX-1)

3: 399-533 feet (DSX-1)

4: 533-655 feet (DSX-1)

5: -7.5db (CSU)

6: -15db (CSU)

7: -22.5db (CSU)
```

If the span is a BRI port, the line build-out is not used and should be set to 0.

framing

one of d4 or esf for T1, or cas or ccs for E1 should be used. Use ccs for BRI. d4 could be referred to as sf or superframe.

coding

one of ami or b8zs for T1, or ami or hdb3 for E1 should be used. Use ami for BRI.

- For E1, there is the optional keyword crc4 to enable CRC4 checking.
- If the keyword yellow follows, the yellow alarm is transmitted when no channel is open.

```
#span=1,0,0,esf,b8zs
#span=2,1,0,esf,b8zs
#span=3,0,0,ccs,hdb3,crc4
```

###

- Here is what the basic entries look like for the B410P card to set ports 1 and 2 to TE mode and ports 3 and 4 to NT mode (remember, each BRI port supports two voice channels).

Take a look at `/etc/dahdi/system.conf` (each line starting with a "#" is a comment):

```
# Span 1: B4/0/1 "B4XXP (PCI) Card 0 Span 1" (MASTER) AMI/CCS
span=1,1,0,ccs,ami
# termtype: te
bchan=1-2
hardhdlc=3
echocanceller=mg2,1-2
# Span 2: B4/0/2 "B4XXP (PCI) Card 0 Span 2" AMI/CCS
span=2,1,0,ccs,ami
# termtype: te
bchan=4-5
hardhdlc=6
echocanceller=mg2,4-5
# Span 3: B4/0/3 "B4XXP (PCI) Card 0 Span 3" AMI/CCS
span=3,0,0,ccs,ami
# termtype: nt
bchan=7-8
hardhdlc=9
echocanceller=mg2,7-8
# Span 4: B4/0/4 "B4XXP (PCI) Card 0 Span 4" AMI/CCS
span=4,0,0,ccs,ami
# termtype: nt
bchan=10-11
hardhdlc=12
echocanceller=mg2,10-11
# Global data
loadzone   = uk
defaultzone      = uk
```

In the above example, the `hardhdlc=x` lines are identifying the data channels on each BRI port.

With these entries in /etc/dahdi/system.conf, when we do a # dahdi_cfg -vvv (command), this is what we should see:

```
[root@localhost dahdi-tools-2.1.0.2]# dahdi_cfg -vvvv
DAHDI Tools Version - 2.1.0.2

DAHDI Version: 2.1.0.3
Echo Canceller(s): MG2
Configuration
======================

SPAN 1: CCS/ AMI Build-out: 0 db (CSU)/0-133 feet (DSX-1)
SPAN 2: CCS/ AMI Build-out: 0 db (CSU)/0-133 feet (DSX-1)
SPAN 3: CCS/ AMI Build-out: 0 db (CSU)/0-133 feet (DSX-1)
SPAN 4: CCS/ AMI Build-out: 0 db (CSU)/0-133 feet (DSX-1)

Channel map:

Channel 01: Clear channel (Default) (Echo Canceler: mg2) (Slaves: 01)
Channel 02: Clear channel (Default) (Echo Canceler: mg2) (Slaves: 02)
Channel 03: Hardware assisted D-channel (Default) (Slaves: 03)
Channel 04: Clear channel (Default) (Echo Canceler: mg2) (Slaves: 04)
Channel 05: Clear channel (Default) (Echo Canceler: mg2) (Slaves: 05)
Channel 06: Hardware assisted D-channel (Default) (Slaves: 06)
Channel 07: Clear channel (Default) (Echo Canceler: mg2) (Slaves: 07)
Channel 08: Clear channel (Default) (Echo Canceler: mg2) (Slaves: 08)
Channel 09: Hardware assisted D-channel (Default) (Slaves: 09)
Channel 10: Clear channel (Default) (Echo Canceler: mg2) (Slaves: 10)
Channel 11: Clear channel (Default) (Echo Canceler: mg2) (Slaves: 11)
Channel 12: Hardware assisted D-channel (Default) (Slaves: 12)

12 channels to configure.

Setting echocan for channel 1 to mg2
Setting echocan for channel 2 to mg2
Setting echocan for channel 4 to mg2
Setting echocan for channel 5 to mg2
Setting echocan for channel 7 to mg2
Setting echocan for channel 8 to mg2
Setting echocan for channel 10 to mg2
Setting echocan for channel 11 to mg2
[root@localhost dahdi-tools-2.1.0.2]# 
```

This has enabled the card in Linux (not Asterisk) and we can check this by running the dahdi_tool:

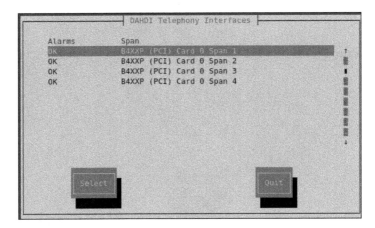

To enable the channels on the card in Asterisk, we need to add some entries in
`/etc/asterisk/chan_dahdi.conf`:

```
[channels]

switchtype=euroisdn      ;identifies the actual signalling system
context=from_outside
signalling=bri_cpe       ;identifies that the following ports are
;acting as the 'user' end of the circuit
group=1
channels=>1-2

group=2
channels=>4-5

context=phones
signalling=bri_net       ;identifies that the following ports are
;acting as the 'network' end
group=3
channels=>7-8
group=4
channels=>10-11
```

In common with the analog channels (because they are all DAHDI), we can dial out
over these by addressing an individual channel or a group of channels.

`exten => _9.,1,Dial(DAHDI/4/${EXTEN:1})` or

`exten => _9.,1,Dial(DAHDI/g1/${EXTEN:1})`

Lastly, let's look at the corresponding entries for the PRI (Digium TE122P) card.

Here are the entries for `/etc/dahdi/system.conf`, when we want to set up an
E1 PRI:

```
span=1,1,0,ccs,hdb3,crc4
bchan=1-15
dchan=16
bchan=17-31
```

When we have done this, we can do a `dahdi_cfg -vvv` and this is the result we should see:

```
[root@localhost ~]# dahdi_cfg -vvvv
DAHDI Tools Version - 2.1.0.2

DAHDI Version: 2.1.0.3
Echo Canceller(s):
Configuration
======================

SPAN 1: CCS/HDB3 Build-out: 0 db (CSU)/0-133 feet (DSX-1)

Channel map:

Channel 01: Clear channel (Default) (Slaves: 01)
Channel 02: Clear channel (Default) (Slaves: 02)
Channel 03: Clear channel (Default) (Slaves: 03)
Channel 04: Clear channel (Default) (Slaves: 04)
Channel 05: Clear channel (Default) (Slaves: 05)
Channel 06: Clear channel (Default) (Slaves: 06)
Channel 07: Clear channel (Default) (Slaves: 07)
Channel 08: Clear channel (Default) (Slaves: 08)
Channel 09: Clear channel (Default) (Slaves: 09)
Channel 10: Clear channel (Default) (Slaves: 10)
Channel 11: Clear channel (Default) (Slaves: 11)
Channel 12: Clear channel (Default) (Slaves: 12)
Channel 13: Clear channel (Default) (Slaves: 13)
Channel 14: Clear channel (Default) (Slaves: 14)
Channel 15: Clear channel (Default) (Slaves: 15)
Channel 16: D-channel (Default) (Slaves: 16)
Channel 17: Clear channel (Default) (Slaves: 17)
Channel 18: Clear channel (Default) (Slaves: 18)
Channel 19: Clear channel (Default) (Slaves: 19)
Channel 20: Clear channel (Default) (Slaves: 20)
Channel 21: Clear channel (Default) (Slaves: 21)
Channel 22: Clear channel (Default) (Slaves: 22)
Channel 23: Clear channel (Default) (Slaves: 23)
Channel 24: Clear channel (Default) (Slaves: 24)
Channel 25: Clear channel (Default) (Slaves: 25)
Channel 26: Clear channel (Default) (Slaves: 26)
Channel 27: Clear channel (Default) (Slaves: 27)
Channel 28: Clear channel (Default) (Slaves: 28)
Channel 29: Clear channel (Default) (Slaves: 29)
Channel 30: Clear channel (Default) (Slaves: 30)
Channel 31: Clear channel (Default) (Slaves: 31)

31 channels to configure.

[root@localhost ~]#
```

A quick look at the dahdi_tool should show something like this:

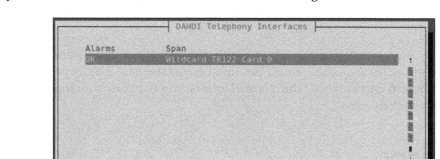

If all is well, you can add the appropriate entries into `/etc/asterisk/chan_dahdi.conf` to make these PRI channels available in Asterisk. There are lots of lines we could put in, but this minimal configuration will get us up and running.

```
[channels]
switchtype=euroisdn
signalling=pri_cpe
context=from_outside
group=1
channels => 1-15
channels => 17-31
```

Again, using the channels within the dialplan is the same as seen in the previous examples—with the PRI channels being treated just like all the others through DAHDI. However, with PRI, groups (and the strategies outlined to use them) are much more useful.

Troubleshooting with Digium cards

Armed with all this information, troubleshooting becomes a fairly binary process.

Assuming the correct driver modules are loaded, if you are encountering problems with lines coming in through a card and you are sure that all physical connections are good, the next step would be a `dahdi_cfg -vvv` command.

If this returns any errors, the issue will be with the card or the configuration in `/etc/dahdi/system.conf`. The `lspci` command will reveal whether the host is seeing the card.

If `dahdi_cfg -vvv` reveals a clean bill of health, the problem is almost undoubtedly in `/etc/asterisk/chan_dahdi.conf`. Common errors here are mainly typos, but it is also worth checking that the channel numbering matches with that set in `/etc/dahdi/system.conf`.

Summary

This chapter has covered traditional telephony interfaces to which you may need to connect Asterisk. Analog, BRI (really only in Europe), and PRI connections are all still alive and well, although the number of new installations of these connections is very much on the decline, as VoIP connectivity begins to take hold.

We have discussed the options that exist to connect Asterisk with these traditional interfaces, which are gateways, external adaptors, or internal cards.

Gateways are usually easier to implement, but do come at the cost of a number of extra physical connections to manage (and therefore potential failure points to consider). However, gateways do provide great building blocks for the creation of redundant and more resilient systems.

Internal cards are generally a neater solution, but require not only their physical installation into the Asterisk PC, but also the installation of Linux kernel drivers before we can connect the channels they provide to Asterisk.

We finished with some example installation using the Digium TDM422 analog card, the Digium B410P BRI (ISDN2) card, and the Digium TE122P PRI card to, respectively, give two FXS (telephone) and two FXO (analog line) connections, four BRI connections, or one PRI connection to Asterisk.

May all your dealings with traditional telephony connections be happy ones.

10
Integrating Asterisk with Wireless Technologies

Of all the chapters in this book, this one is probably the one that will mention Asterisk the least! And yet, there is good reason for its inclusion—over a number of years of meeting people who create and deploy Asterisk-based solutions, I have found that they do not seem to be aware of the different options for hooking up Asterisk to the wireless world.

Most people have played with one or two of the options, but do not seem to have been exposed to the many ways that exist. Furthermore, the majority of those I meet that have attempted to connect Asterisk with a wireless device (usually a Nokia dual-mode GSM/Wi-Fi SIP phone), have given up after a number of attempts to find the right settings, both on the device and in Asterisk, that achieve a workable hook-up.

Therefore, the idea of this chapter is to give an introduction to the area of wireless technologies by asking, "Why integrate Asterisk with wireless technologies?"

After answering that question, we will look at the wireless device and wireless network options that exist, and consider the advantages and disadvantages for each. We will also look at some configurations for one or two devices and the settings we need to make in Asterisk, before rounding off the chapter with some example deployment scenarios, for which we will choose the best wireless options.

How does that sound for this chapter? Good? So sit back, relax, and enjoy the ride!

Why integrate Asterisk with wireless technologies?

This is a good question, and is worthy of consideration at the outset of the chapter. As an open-standards (never mind open source) telephony platform, Asterisk is ideally positioned to connect with all manner of devices, and this is excellent news for those that are involved in designing and deploying Asterisk-based solutions, as we are able to choose the best options rather than being restricted by proprietary compatibility issues.

Getting back to the question, or rather the answer to the question, mobility is one of the main reasons people want to hook Asterisk up to the wireless world—both mobility within the office environment and the requirements of a mobile workforce outside the office. Even in this day and age, many international travelers are plagued by heavy roaming costs for their mobile phones. Ironically, a good proportion of them probably carry devices with Wi-Fi and VoIP capabilities, some even staying in hotels with free Wi-Fi, if only they knew!

Other reasons that people and corporations may be looking for wireless solutions include rapid deployments, cost reductions (in implementation costs, running costs, or both), or it may be that they need a portable PBX solution that can be moved around with them without the hassle of running cables and so on, every time they arrive at a new location.

The following table shows business drivers and technology enablers involved:

Business drivers	Technology enablers (and issues)
<ul><li>Mobility within the office<ul><li>○ Work anywhere in or around the building</li></ul></li><li>Mobile workforce<ul><li>○ International travel</li><li>○ Field staff</li><li>○ Multi-site enterprises</li></ul></li><li>Rapid implementations<ul><li>○ New office installations*</li><li>○ Fast office installation</li></ul></li></ul>	<ul><li>Wireless access points are already very common, both within offices and in public locations</li><li>Wi-Fi and WiMAX technology (may suffer with NAT and firewall issues)</li><li>Wireless routing means phones (and PCs) can now be rolled out very quickly</li><li>SIP and IAX2 trunks are usually much more cost effective that traditional trunks</li></ul>

Business drivers	Technology enablers (and issues)

• Cost reductions
   ◦ Reduced "hard" infrastructure (cabling, switches, and so on)
   ◦ Reduced real-estate needed
• Temporary deployments
• Hot desking
• Business continuity*
• Disaster recovery*
	• IP PBXs can be physically much smaller, meaning less rack/floor space is used
	• The power consumption of an IP PBX can be a lot less than a traditional PBX.

*These reasons in particular are a great sell for VoIP, not just wireless.

Wireless technology overview

With this weight of reasons to think about wireless, our next job is to look at the options available so that we can deploy the best technology in each given application. To help us consider the options, I have split the next section into two parts—one that considers wireless handsets and one that considers wireless networks.

There are many types of wireless handsets on the market today, some have been around for quite some time while others are relatively new. Most are aimed at the consumer, a few are definite enterprise plays, and there is even an "industrial strength" offering. Let's outline each type and home in on their advantages and disadvantages.

Wi-Fi (only) phones

Wi-Fi (only) phones have probably been around the longest, but in most cases, the vendors do not seem to have refined their offerings following the experience of their first products. Rather, Wi-Fi (only) handsets seem to be disappearing in favor of dual-mode cell phones. These devices started out as a piece of Wi-Fi kit that someone attached to a SIP stack and added on a microphone, speaker, and some basic audio-processing apparatus—and it shows.

These phones are usually not excellent. Their issues often include poor battery life and less than perfect audio. Also, something that could be a big issue depending on your specific application, is the ability to roam from one Wi-Fi hotspot to another, which has now been addressed in specification 802.11n. But I have not yet seen many phones which implement this advance.

Advantages	Disadvantages

• Truly mobile
  ◦ Inside the office or home
  ◦ Other hotspots worldwide
• Allows PBX extension to travel
• Small and portable
	• Configuration is difficult
	• Battery life is generally short
	• Voice quality can be poor
	• Not a wide choice of phones
	• If the phone does not have a web browser, you may not be able to connect to hotspots that require a login

One notable exception to majority of Wi-Fi (only) handsets is the Polycom range. As you would expect, these phones are very well constructed — there is even a rough use version for demanding environments, and they work well too.

For more information, look up UT Starcomm, Hitachi, or Linksys Wi-Fi phone.

SIP desk phones with a wireless link

Imagine all the positive attributes of a good SIP "normal" desk phone — great looks, familiar feel, excellent functionality, high speech quality, and sturdy construction.

Now make it wireless!

This is exactly the idea that Linksys (and others like Mitel) have implemented by bringing out a wireless Ethernet bridge (a device that connects into a standard Ethernet socket on any device and gives it Wi-Fi connectivity), which is specifically designed to fit into the void in the base of their phones.

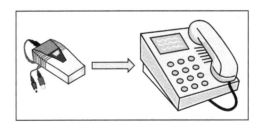

This brings together the standard-looking office phone (the sort that seems familiar to users and will not scare them) with the convenience of wireless. The intention is not to go wandering around with this phone, as it still needs mains power, but to be able to locate it wherever you want in the office without needing a cabled Ethernet connection, nice!

Advantages	Disadvantages
• Very user friendly • Great for office use • Comprehensive features • Good for temporary and flexible deployments	• Additional work of configuring and connecting the wireless Ethernet bridge • The devices still need power, and so are not truly wireless • Adds significant extra cost per extension

Of course, any wireless Ethernet bridge and any SIP desk phone could be used, as these are standard interfaces that we are talking about. You could just have a wireless Ethernet bridge for a whole room going into a switch and then run cables to a number of phones (if you needed a few), but the elegance and simplicity of the Linksys solution does make it stand out.

Dual-mode (GSM and SIP) phones and PDA/smart phones

Within the last two to three years, these devices have really taken off. There are offerings from UT Starcomm (who also make Wi-Fi [only] handsets) and Pirelli Communications, but it is the Nokia handsets that have set the standard.

Any high end "E" (business) or "N" (multimedia) series Nokia handset will have both Wi-Fi and SIP connectivity, and a SIP client is available for the iPhone, and for Windows mobile phones and PDA devices too. Great news for those wishing to integrate with Asterisk!

These handsets (putting call costs to one side for a moment) provide the ultimate mobility, as they will enable calling through Wi-Fi access points or over the regular mobile network.

Often held up as examples of **FMC (Fixed Mobile Convergence)**, GSM/SIP phones are *not really* that—they are really two phones in one case, a GSM (or 3G) phone and a SIP phone. The only converged thing about them is that they share a microphone, an earpiece, and the contacts list. These devices cannot currently be configured to intelligently route calls over GSM or IP, depending on, say, cost. You have to choose your preferred method of communication, and all calls will go that way unless you choose the alternative (with an inconvenient two key-press method) on a per-call basis.

This is something of an issue in countries like the UK, where it is usually cheapest to call a mobile from a mobile, and most business customers will have their mobile device on a monthly plan that includes minutes—so they will want calls to mobiles routed over the mobile network, and calls to landlines routed over the SIP (Wi-Fi) network.

Having said that, two things have emerged to make things a little more user-friendly.

Firstly, the UT Starcomm phones have two "place call" buttons—one that directs the call over the GSM (or 3G) network and one that directs the call over the SIP (Wi-Fi) network. Secondly, some third-party companies have developed small applications which run on the Nokia handsets (which use the Symbian S60 operating system) to do Least Cost Routing, which to my mind is an excellent development.

Those little gripes aside, can you see the power of a single handset which is both a standard mobile phone AND an extension of your Asterisk PBX, by virtue of its SIP and Wi-Fi capabilities? To give you an idea of the potential benefits, co-author David Duffett describes his experiences:

> *As someone who travels regularly, I continue to be impressed when my Nokia phone rings while I am in, say, Johannesburg, South Africa because someone back in the UK called my office number! It freaks them out when they find out you are not really in the office, and they are also impressed when you tell them that the call between your telephone system (Asterisk) and your mobile phone is FREE! It is heart-warming to find yourself in a hotel that offers free WiFi, able to make calls back to other extensions in the office at zero cost. And, as the internet (in general) and WiFi access points improve, I am finding that there is no noticeable difference in quality between a regular mobile call and a call placed over the internet.*
>
> *I have recently been experimenting with placing SIP calls over the 3G network (as opposed to using WiFi internet access) and, where the 3G network is good (mainly town and city centers at present), call quality has been acceptable. This means that if you are on an unlimited data plan as part of your mobile package, you could be calling your office (and any locations peering with it) free.*

Advantages	Disadvantages
• Typically better voice quality than Wi-Fi (only) phones	• Battery life is reduced by having Wi-Fi on, in addition to the regular phone
• User interface is generally better than Wi-Fi (only) alternatives	• Complex hotspot attachment processes in some environments (not a problem with the phone, just the Wi-Fi hook up)
• Great mobility	
• Always on-net, just variable costs	
• VoIP is an "application"	• Sometimes frustrating steps to choose call route
• Respected vendors	

For more information, look at Nokia, Pirelli, or UT Starcomm dual-mode VoIP phone, or Windows mobile SIP.

SIP/DECT phones

A world away from the poor speech quality and short battery life that plagues most Wi-Fi (only) phones are good old DECT cordless phones. They have been around for years and are an optimized technology.

A healthy DECT handset will usually last for days (not hours) on a single charge and the speech quality is excellent, unless you venture to the very edge of its range, which incidentally, is much greater than that of a Wi-Fi (only) phone.

Some clever people thought—Why don't we harness the excellence of the DECT air interface (or radio link) with the ubiquity of SIP for IP telephony, by putting a SIP stack in the DECT base station? So they did, and the results are amazing.

Now, don't expect to take your DECT handset to the airport and make calls from there. This won't work—if you want that kind of functionality, it has to be the Wi-Fi (only) or dual-mode phone for you. But, for in-office mobility, these SIP/DECT phones really are fantastic.

Integration with Asterisk is very easy, it's just a SIP connection. Do note that some of these are consumer offerings, and as such, they have an analog connection on the base station in addition to the LAN connector—the idea being that you connect both and choose which way to route calls by pressing the appropriate buttons on the phone.

Since we are interested in Asterisk installation, we will not use the analog connection and all calls will be routed to Asterisk via SIP.

It is usual to find that multiple handsets can be attached to a base station, and each base station can handle a number of SIP accounts, but beware that each base station will have a restriction on the number of concurrent calls that it can handle, which may be as low as two!

Make sure that you know the limitations of the base stations you are using, and that you deploy the number of base stations you need in order to facilitate the volume of concurrent calls you want. Each handset is linked with a SIP account (or accounts) by checking boxes on the web interface for the base station.

It should also be noted that there are several enterprise-grade SIP/DECT solutions (notably Aastra and Polycom), which offer further facilities including better capacities per base station and much better roaming abilities; some even allow multi-site roaming.

Advantages	Disadvantages
• Excellent range when compared with Wi-Fi • Great battery life • Very good speech quality • Intuitive use (as in the case of a standard cordless phone) • Standard SIP configuration	• Only mobile within the office • Linking handsets with SIP accounts on the base station web interface can be a little confusing the first time you do it

The preceding four sections represent the different types of wireless handsets that can be connected to Asterisk. Any wireless solution you propose is going to be made up of one or more of these options.

To conclude this section of the chapter, here is a table comparing the four handsets:

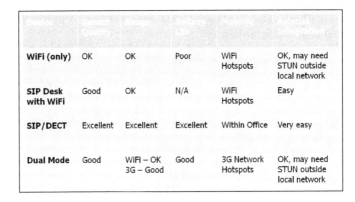

WiFi (only)	OK	OK	Poor	WiFi Hotspots	OK, may need STUN outside local network
SIP Desk with WiFi	Good	OK	N/A	WiFi Hotspots	Easy
SIP/DECT	Excellent	Excellent	Excellent	Within Office	Very easy
Dual Mode	Good	WiFi – OK 3G – Good	Good	3G Network Hotspots	OK, may need STUN outside local network

Connecting Asterisk to mobile networks

There are times when you will want to connect to Asterisk to a wireless network. This can range from the simplicity of adding a standard wireless access point to the Ethernet network (so that wireless devices can connect to the network, and therefore to Asterisk), to the somewhat more esoteric area of giving Asterisk a direct connection to a mobile network. As the addition of a wireless access point is really standard networking practice, and not specific to VoIP, it will not be covered in this chapter.

Why connect to mobile networks?

Before we look at the two options for connecting Asterisk directly to mobile networks, let's examine the reason for doing it, which is usually "cost".

In a lot of countries, it is fixed landline to mobile calls which are the most expensive and this is due to the interconnect charges made by the mobile operators to the landline telco, and the profit margin of that telco. The following diagram depicts this situation:

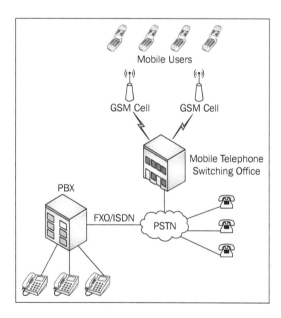

This high-cost situation is in sharp contrast to rates of mobile to mobile calls, which are often included in mobile charging plans (for example, 500 minutes per month), or are sometimes totally free between SIMs on the same mobile network.

This fact has not been left unnoticed by the creative telephony community, and for some time now, devices called GSM gateways have been available. These devices usually have an external aerial and a cut down mobile phone circuit into which you install a SIM for access to a given mobile network. Until recently, the connection to the PBX was an analog line, which would work with Asterisk, but would need the introduction of an analog interface in addition to the GSM gateway.

More recently, not only have these GSM gateways been enhanced to give direct SIP connectivity, but another hardware option has emerged in the form of GSM cards that are installed in the PC. Here is a revised version of that previous diagram, showing the most cost-effective way of calling mobiles from the PBX. Although this diagram shows a GSM gateway box, the principle applies to the GSM card implementation too.

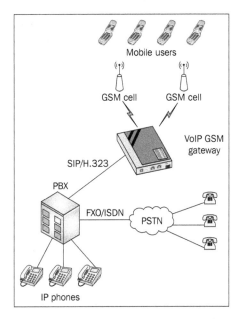

Having looked at the concept, we can now evaluate the two types of implementation.

The GSM gateway (box)

Although the concept of the GSM gateway box has been explained previously, the actual equipment available varies from single SIM to multiple SIMs, connecting via an FXO line, PRI connections, or SIP (there may even by some H.323 models out there).

Here are some examples:

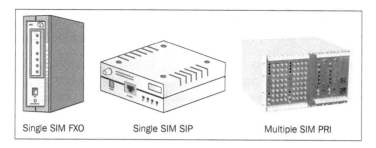

Single SIM FXO   Single SIM SIP   Multiple SIM PRI

Because these units are designed to use up the "free" minutes included within the monthly plan for the SIMs, it is common to find more SIMs than radio circuits. The presence of this situation means that there is undoubtedly embedded software somewhere that allows the user to enter the number of included minutes for each SIM, and then, when the unit determines that the 500 minutes, say, from SIM "A" have been used up, it will seamlessly switch over to SIM "B".

Of course, to get a reasonable return on investments for such (multi SIM) systems really requires that your customer currently has quite a heavy fixed-to-mobile call volume.

Even at the single user end of the market, money can be saved, and not only by routing outbound fixed-to-mobile calls straight into the mobile network using one of these (single SIM) devices. Keep an eye out for my "neat money saving tricks" further in this chapter.

Advantages	Disadvantages

• Not platform dependent
• Connect via FXO, SIP, or PRI for large scale deployments
• Easily scalable—just add more boxes
• No special drivers required
	• Lots of wires
	  ◦ Antenna
	  ◦ Power
	  ◦ Connection to Asterisk
	• Care needed to ensure inclusive minutes are used effectively

The GSM card

The alternative to a GSM gateway box is a GSM card. PCI cards are still the dominant type available, but there are sure to be PCIe versions around too, which will become dominant as the PCs of today have either one or no PCI slots.

Even when installing a PCI card into a PCI slot in a PC, there can be issues such as minor bus incompatibilities (which can have a major effect on functionality), IRQ issues, and so on. Fortunately, most cards are compatible with both the 5.0V and 3.3V PCI slots found in PCs, so that is one less issue to worry about.

Although a number of cards have the same number of SIM slots as radio circuits (as indicated by the number of antenna sockets on the cards), it is possible to get cards with more SIM slots than radio circuits, for the reasons outlined in the section.

Cards are the "neater" solution, as the only cable involved is the antenna cable, there are no power supplies or Asterisk connections (usually FXO, PRI, or SIP) that you need to deal with, as all of that is looked after by the host PC platform.

Of course, these card solutions do come with the need to install drivers to get working within the Linux environment before you can make use of them within Asterisk. So do make sure that any cards you purchase have a good reputation for ease of installation.

Advantages	Disadvantages
• Neat solution, card is inside the PBX platform • Less interconnecting cables, just the antenna	• Platform compatibility is essential • Expansion may prove difficult, as slots for further cards are required • Setup and configuration can be difficult • Routing coaxial antenna cables (when necessary) could prove problematic

Configuring wireless devices

Each device will have its own user interface—either through the display on the device (if it's a phone) or through a web interface that you can browse to (some Wi-Fi phones will have both).

The actual entries used for setting up the device to work with Asterisk are standard SIP parameters like username, password, host, and so on, and therefore, I have included the next screenshot (taken from the Siemens S460IP phone) as an example:

IP Configuration	SIP		?
Telephony	Authentication Name:	205	
Connections	Authentication password:	••••	
Dialing Plans	Confirm authentication password:	••••	
Advanced Settings	Username:	205	
Miscellaneous	Domain:	192.168.2.200	
	Display name:	205	
	Proxy server address:	192.168.2.200	
	Proxy server port:	5060	
	Registrar server:	192.168.2.200	
	Registrar server port:	5060	

Network

STUN enabled:	○ Yes ● No	
STUN server:		
STUN port:	3478	
NAT refresh time:	20	sec
Outbound proxy mode:	○ Always ● Auto ○ Never	
Outbound proxy:	192.168.2.200	
Outbound proxy port:	5060	

Voice codecs

VoIP Volume: ○ Low ○ Normal ● High

Enable Annex B for G729: ○ Yes ● No

Selected codecs		Available codecs
G711 a law / G711 µ law	< Add / Remove > / Up / Down	G726 / G729

What is worth looking into in more detail are the Nokia dual-mode phones, as there are a couple of parameters necessary for successful connection with Asterisk, which are not as widely known as they should be. These "lesser known" parameters are responsible for many people giving up on connecting their devices to Asterisk.

I will use the Nokia E90 as an example:

Most of the parameters are self explanatory, but the two parameters we pay special attention to are the **Public user name** in the SIP profile itself, and the **Realm** in the submenus for both the **Proxy server** and the **Registrar server**. Further in this section, I have included screenshots of each of the screens you will need to go through to set up your Nokia phone as an extension on your Asterisk PBX.

Start by selecting the **Connection** in the **Settings** screen:

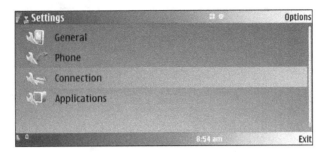

Then select **SIP settings:**

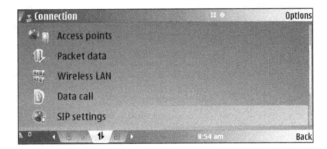

Select or create the relevant SIP profile from the submenu:

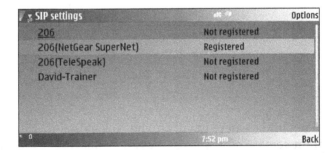

Inside the SIP profile, complete the settings — the **Public user name** must be the SIP profile name (from `sip.conf`)@<the IP address of your Asterisk box>:

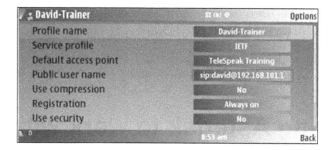

Now move down to the **Proxy server** and **Registrar server** sections:

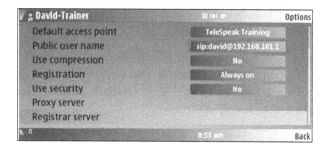

It is here that the all-important **Realm** *must* be entered — Asterisk sets this as "asterisk" by default, but you can change it in the `[general]` section of `sip.conf`.

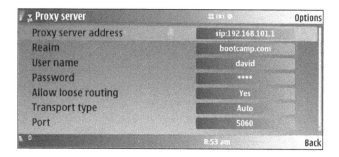

The **Registrar server** screen needs to be the same as the previous one:

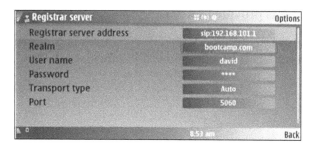

Although the IP address used (see preceding screenshot) is internal, it could equally be an Internet IP address such as 80.229.203.112

Note that the **Realm** is defined in the [general] section of sip.conf by the realm= line. If this line is missing, then the realm will default to "asterisk".

Configuring Asterisk to work with wireless technologies

This is going to be quite a short section. It is very easy to configure Asterisk to integrate with most wireless devices because they are just standard SIP devices; you don't need to learn any new stuff!

If you plan to use any devices (including wireless) outside of your own network domain, which will be true for a lot of cases, you will need to tell the Asterisk SIP channel about this—in terms of the **NAT** (**Network address translation**) arrangement and your external IP address.

Before we do that, here is a diagram to show why this is necessary.

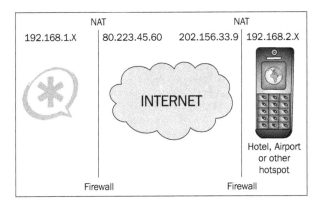

Taking the Asterisk side of the diagram, the internal domain is 192.168.1.X, but the address our broadband router (or whatever) uses on the outside (the Internet) is different. This address is provided by our ISP.

We need to tell Asterisk about this, so that it recognizes both the internal addresses and the external address in the transactions it handles.

Here is how this is done in `sip.conf`

```
[general]
context = default
realm = asterisk
externip = 80.229.109.204
localnet = 192.168.2.0/255.255.255.0
bindport=5060
srvlookup=yes
disallow=all
allow=ulaw
allow=gsm
language=en
```

The key lines are the `externip` line (telling Asterisk the Internet IP address) and the `localnet` line (telling Asterisk the internal address range).

If you have a resolvable URL and access to DNS, then you can use `externhost` instead, for example,

```
externhost=david.dyndns.org
```

Once you have let Asterisk know this information in the `[general]` section, then you just need to identify the individual endpoint(s) that will be used on the other side of the NAT on the right (which is your mobile or remote devices) by using the `nat=` line in their SIP profiles.

```
[my_mobile_device]
type=friend
host=dynamic
secret=1234
context=default
...
...
nat=yes
qualify=yes
```

The `nat=yes` line tells Asterisk to disregard the internal (192.168.2.x) address the remote device will give during SIP interactions in favor of the external (202.156.33.9) address by which the device will be recognized on the Internet. You will notice that, in addition to the `nat=yes` line, I have also included a `qualify=yes` line. This line tells Asterisk to periodically check with the device to ensure it is on the network. By using a figure (for example, `qualify=200`) instead of `yes`, we can instruct Asterisk not only to check on the device, but also to treat it as unavailable if the latency between it and Asterisk goes above 200 ms (in this case).

Deployment choices

The type of devices or combination of devices you choose for a particular implementation will depend on a number of factors and each job will be different, but I thought it would be useful to look at a few deployment scenarios together with the kind of device which may be best suited for them.

- Office only
  - SIP/DECT (range, battery life, speech quality)

- Mobile
  - Wi-Fi (only) if no 3G/GSM requirement
  - Dual-mode (for example, SIP client on all Nokia E phones)
  - PDA with SIP or IAX soft client

- Remote
  - Laptop with SIP or IAX soft client

- Least Cost Routing (mobile calls over the mobile network)
  - Gateway box
    - Easy SIP integration
    - No worries about "slot" compatibility
    - A range of boxes, easy to scale
  - Gateway card
    - Neat—no other boxes or PSUs
    - up to four SIMs per card

- Redundancy (successfully routing calls in the event of a landline failure)
  - Gateway box

Neat money saving tricks

Just before we end this chapter with a summary, I thought it would be useful to show you a couple of money saving tricks, which all fall under the category of **least cost routing** (or **LCR**), but are perhaps a little more creative.

Calling a mobile phone

If you have a dual-mode phone as an extension of Asterisk, always try to route the call over SIP first (in case the handset is within Wi-Fi coverage), even if the caller dials the standard mobile number. Then, if you do need to route the call to the mobile network, do it directly with a GSM gateway. When that fails, if you must go via the fixed network, can you route the call via an alternative carrier to reduce the cost of the call? Only after you have tried those, would you look to route the call in the "normal" way.

In the UK-based example that follows, the number starting with "077" is the standard mobile number, and there is a carrier called "18185" (www.18185.co.uk), which, at the time of writing, provided the cheapest way to call a UK mobile phone from a UK fixed line. This dual-mode phone is also set up as endpoint 2001 of your Asterisk PBX.

Here is what your dialplan might look like:

```
exten => 07711223344,1,Dial(SIP/2001,30)
exten => 07711223344,n(gsm-gateway),Dial(SIP/gsm-gateway/${EXTEN})
exten => 07711223344,n(alternative),Dial(DAHDI/1/18185${EXTEN})
exten => 07711223344,n(pstn),Dial(DAHDI/1/${EXTEN})
```

Avoiding those nasty roaming charges that arise from receiving calls

One of the worst kinds of roaming charges (other than mobile data roaming, which seems to attract the mother of all sky-high rates), are the charges you incur when receiving calls while you are travelling out of your home country.

Different countries have different mobile charging plans, so the following example may not directly apply to your scenario, but you will see the principle, and you may be able to use it in some other way.

In the UK, there is no charge for receiving a voice call on your mobile phone in your home network, but the high level of the charges for receiving a call while abroad (when you pay for the international leg of the call) can only be described as wicked! Let's take a situation where someone from the UK is going on a two-week business trip to the US. They either have a dual-mode phone or soft client (on a laptop or PDA), which is set up as SIP endpoint 2005. Knowing that receiving calls over GSM is going to be very expensive, they take the SIM out of their phone and put it into a GSM gateway connected to their Asterisk PBX in the UK. Now, all they have to do is try to call their SIP extension when a call comes in on their mobile SIM, and, wonder of wonders, NO roaming charges!

Of course, this system does rely on them being in a Wi-Fi hotspot, with their dual-mode phone or soft client ready to take the call. If they are not, we could send the call to voicemail so that they can collect it later. Or we could also call the US number that they are available on (their hotel or office, or a US pay-as-you-go mobile) — as calls to the US from the UK are pretty cheap, especially over a SIP trunk provider. Here is the example dialplan (assuming that the SIP call coming in from the GSM gateway gets dropped into the [from-my-sim] context with extension s):

```
[from-my-sim]
exten => s,1,Dial(SIP/2005&SIP/my-trunk/0012564286000,20)
exten => s,n,VoiceMail(2005@default)
```

Summary

The main thrust of this chapter was to acquaint you with the many options that exist to connect Asterisk with the wireless world.

The main options for handsets are Wi-Fi (only), wireless-enabled desk phones, dual-mode phones, and SIP/DECT setups.

We can connect Asterisk to wireless networks like the GSM network by using a gateway box or a GSM card installed in the Asterisk platform itself, similar options may be available for CDMA networks and others.

Armed with this knowledge, and the power and flexibility of the Asterisk dialplan, you can now be very creative — not just in terms of least cost routing, but in tailoring the perfect solution for your customer for any scenario. The limitations are few and the possibilities are great, so go for it!

11
Graphical User Interfaces

This chapter explores the use of **Graphical User Interface**(s) (**GUIs**) for Asterisk. We look at why you might want to consider an Asterisk system with a GUI, and what the good and bad points of such systems are. We also look at a popular, completely open-source GUI — FreePBX, in a bit more depth. As a result, you should be equipped with the knowledge to decide if a GUI-based system is appropriate for your needs or those of your customer.

Reasons for going GUI

One of the major developments in the evolution of computer systems is the introduction of relatively complex user interfaces. These developments were driven by the desire to input information and output feedback in as effective a way as possible through the development of a relatively friendly interface. The wide range of sizes of computer systems found in the present day, from embedded chips to room-filling supercomputers, has resulted in quite a disparity in the forms of human interfaces that have been developed. Those with human interaction as a primary raison d'etre tend to have highly advanced graphical interfaces — PCs, laptops, and mobile phones are obvious examples. Those, where immediate human-oriented input and/or output is less important or very restricted in nature, such as simplistic embedded chips or large number crunchers, tend to have less evolved interfaces.

A number of systems have ended up with multiple human interfaces, most usually those where human interaction is a prime concern. Personal computers are a prime example, where initially all input and output was via a keyboard and screen respectively, but now a whole plethora of channels exist for the transfer of information, including input devices such as mice, microphones, scanners, touch screens, and gaming devices (joysticks and others). The output devices include speakers, printers, and the gaming devices mentioned previously, or at least those with force-feedback capabilities.

Where does an Asterisk PBX sit in the range of systems? Well, it's true to say that its primary function is human interaction, but it is equally true to say that the mechanism for enabling that interaction is predominantly the telephone handset. In common with all PBXs, Asterisk allows relatively complex operations to be carried out through the keypad and handset of a common phone. However, there is scope for some of those operations to be carried out using other devices, a common example being listening to voicemail using the speakers on a PC. There are also administrative and maintenance operations that require manipulation of the configuration files on the Asterisk server, from something as simple as adding a new extension to something more complex such as prioritizing calls in a range of call queues. Again, the usual mechanism for this is either directly on the server or via a PC on the LAN.

GUIs for Asterisk almost exclusively work to ease the burden of administration and enhance the options available to the end user, and exist in a number of forms. For instance, there is a software package **ARI (Asterisk Recording Interface)**, that you can install on an Asterisk server to allow users to access their voicemail using their PC. In this chapter, we will focus on graphical interfaces that have "server administration" as their primary goal. While other GUIs are pretty much accepted across the board, the choice of whether or not to administer your server in this fashion is one that generates much discussion, and can potentially have the biggest impact on your system.

Good to GUI

Depending on your needs or your customer's particular needs, there can be significant advantages to choosing an Asterisk system that has an administration GUI. In practice, these boil down to:

- Ease of administration
- Access to enhanced features
- Easier upgrade process
- Standardized code

Ease of administration

The most obvious reason for a company choosing an Asterisk system with a graphical interface over a system without one, is ease of administration. As a result, all major PBXs on the market now come with graphical administration interfaces. With a "vanilla" Asterisk system, a simple administrative task such as adding an extension is not onerous, involving, as it does, the addition of a few lines of text to `extensions.conf`, `voicemail.conf`, and `sip.conf` (or `iax.conf` if you have

IAX-capable phones), but it is a task that should really be carried out by someone familiar with Asterisk configuration in general, who is also comfortable with the Linux command line interface. Let someone without those competencies loose on your Asterisk box, and you run the risk that, in extreme circumstances, a typo could incapacitate the system until a third party gains physical access to fix the problem. It's a significant risk, given that one of the primary goals of a PBX is to "just work". Frequently, that risk is mitigated through the use of a support contract with a third party that burdens them with the responsibility for making all those minor amendments. Such a contract obviously involves an ongoing support cost, but also requires the implementation of a secure mechanism to give the third party access to the PBX, but ideally not the whole network. There is also the need to introduce an effective process for requesting changes internally. You certainly don't want every user having the authority to call the third party directly and request any change they fancy.

With a GUI, it is less of a risk for day-to-day admin tasks to be carried out in-house. Indeed, many GUI-based frontend Asterisk systems are managed on a day-to-day basis by office administrators rather than IT specialists, as they can add, move, or delete extensions and manage ring groups without the prospect of typos and with restricted access to the overall system. The better GUIs offer a granular level of access too, so that the simple day-to-day tasks are the only options available to the people with those responsibilities, and the more complex but occasional changes in configuration can only be carried out by a named administrator.

Access to enhanced features

For someone completely new to Asterisk, there is quite a short but steep learning curve to negotiate before they gain the necessary skills and knowledge to set up even the simplest PBX. Progressing to more complex systems can require a significant amount of further investment in the person or people tasked with enabling and maintaining the new functions.

Many Asterisk GUIs do this work for you, and present you with an interface where you have to do little more than make the relevant choices from drop-down menus. Indeed, FreePBX (we'll look at it in more detail later in the chapter) enables advanced features in a modular fashion. You have to choose to install a module for call queues, for example, before you can use this function in your system. This has the advantage of keeping the administration interface as simple and clean as possible. You also end up with well structured and commented code in your dialplan, albeit rather a lot of it.

Of course, this can lead to situations where new functions are implemented because they can be, and not because there is a real need. But this is not the fault of the GUI and the risk can be reduced through the use of robust change control processes.

Easier upgrade process

Upgrading your Asterisk PBX is a topic that can generate very strong opinions on either side of the imaginary fence. It is certainly difficult to justify upgrading the core Asterisk engine just for the sake of keeping up with the latest version. Telephone systems, even those based on PC architecture, follow their own rules in this respect, and right at the head of that rules list is the mantra, "If it ain't broke, then don't fix it". However, there are occasions when it makes sense to upgrade. Maybe you have fallen foul of a bug, or there is a new feature that will allow you to implement much needed functionality. Maybe a vulnerability in the version of Asterisk you are running has been discovered, and there is no prospect of a patch being issued. Whatever the reason, once you've decided to upgrade, you will be faced with ensuring that not only will the new/updated features work as desired, but also that the myriad of features you have already implemented and constantly use in the current release will work.

Like most systems, Asterisk tries to maintain backwards compatibility, meaning that code in one release will work in exactly the same way in a later release. Although, in common with most systems, this is not always achieved. There have been occasions where functions in Asterisk change from one release to the next, usually in the syntax of the code, but sometimes functions are completely deprecated and replaced with an alternative. This necessitates a round of code checking, amendment, and stringent testing.

Using a GUI-frontend system can ease some of that burden. Rather than needing to understand exactly what has changed from one release to the next, and what code needs to be targeted to ensure it still works, the GUI developer will have already gone through that process. They will understand which features available through the GUI will be affected, what code changes are required to maintain functionality, and should have a mechanism for automating those changes. Thus, you will be left to update any custom code you have inserted, and then carry out all the tests you would have done on a "vanilla" system. It's still a significant exercise, but at least some of the pain will have been removed.

Standardized code

Asterisk dialplans, in common with other programming environments, require careful management of code if you are not going to end up with an unholy mess. Liberal use of comments and structured programming techniques (such as using `#include` for standard "subroutines") are highly recommended, particularly if more than one person will be maintaining the dialplan.

However, using a GUI ensures that the majority of the code produced will be very structured, and having been tested by many developers, beta testers, and end users, there is a high degree of confidence that it will do what is expected of it without any "gotchas".

GUI, phooey!

For all the positive reasons for using a GUI, there are also a number of arguments against going down that route. In general, they relate to the added complexity of a system running a graphical interface and fall under the following headings:

- Performance
- Stability
- Restricted functions

Performance

In order to run a graphical interface, the Linux server at the core of each Asterisk system is required to install and run a number of extra packages. Many GUI implementations are based on a so called **LAMPA** architecture, standing for Linux, Apache, MySQL, PHP, Asterisk. Of these packages, only the Linux core and Asterisk are absolutely needed to implement the telephony functionality, the rest is needed to enable the GUI. This requires extra memory, disk and processing power of the host PC/server, which is not a huge concern if you are implementing a standalone PBX for a relatively small office, but if you are rolling out a system for thousands of users over dozens of sites, then all those costs start adding up. Although, in the real world, many Asterisk boxes without a GUI will still run Apache, MySQL, and PHP to allow them to store call data effectively and use **AGI** (**Asterisk Gateway Interface**). This obviously narrows the performance gap between GUI and non-GUI.

Even in a single-site scenario, if the number of users and call throughput is high, then, by choosing a GUI-based system in preference to one without a GUI, you can end up with a more complex and expensive system. For instance, in a call centre, a "vanilla" PBX with reasonable hardware spec will cater for most requirements, as it is not unusual for such a setup to be capable of handling well over 100 concurrent calls. With a GUI-based system, you may need to implement two or more servers to achieve the same goal. You then need to consider which server(s) handle inbound calls, how calls are handed off between servers, and how to cater for the failure of a server. In short, you will probably need to implement server clusters earlier with a GUI-based setup.

Another performance hit of GUI-based systems comes from the use of rather complex code in the dialplan. Let's look at the code produced by FreePBX for initiating an outbound call as an example.

```
[macro-dialout-trunk]
include => macro-dialout-trunk-custom
exten => s,1,Set(DIAL_TRUNK=${ARG1})
exten => s,n,ExecIf($[$["${ARG3}" != ""] & $["${DB(AMPUSER/${AMPUSER}/
pinless)}" != "NOPASSWD"]],Authenticate,${ARG3})
exten => s,n,GotoIf($["x${OUTDISABLE_${DIAL_TRUNK}}" =
"xon"]?disabletrunk,1)
exten => s,n,Set(DIAL_NUMBER=${ARG2})
exten => s,n,Set(DIAL_TRUNK_OPTIONS=${DIAL_OPTIONS})
exten => s,n,Set(GROUP()=OUT_${DIAL_TRUNK})
exten => s,n,GotoIf($["${OUTMAXCHANS_${DIAL_TRUNK}}foo" =
"foo"]?nomax)
exten => s,n,GotoIf($[ ${GROUP_COUNT(OUT_${DIAL_TRUNK})} >
${OUTMAXCHANS_${DIAL_TRUNK}} ]?chanfull)
exten => s,n(nomax),GotoIf($["${INTRACOMPANYROUTE}" =
"YES"]?skipoutcid)
exten => s,n,Set(DIAL_TRUNK_OPTIONS=${TRUNK_OPTIONS})
exten => s,n,Macro(outbound-callerid,${DIAL_TRUNK})
exten => s,n(skipoutcid),AGI(fixlocalprefix)
exten => s,n,Set(OUTNUM=${OUTPREFIX_${DIAL_TRUNK}}${DIAL_NUMBER})
exten => s,n,Set(custom=${CUT(OUT_${DIAL_TRUNK},:,1)})
exten => s,n,GotoIf($[$["${MOHCLASS}" = "default"] |
$["foo${MOHCLASS}" = "foo"]]?gocall)
exten => s,n,Set(DIAL_TRUNK_OPTIONS=M(setmusic^${MOHCLASS})${DIAL_
TRUNK_OPTIONS})
exten => s,n(gocall),Macro(dialout-trunk-predial-hook,)
exten => s,n,GotoIf($["${PREDIAL_HOOK_RET}" = "BYPASS"]?bypass,1)
exten => s,n,GotoIf($["${custom}" = "AMP"]?customtrunk)
exten => s,n,Dial(${OUT_${DIAL_TRUNK}}/${OUTNUM},300,${DIAL_TRUNK_
OPTIONS})
exten => s,n,Goto(s-${DIALSTATUS},1)
exten => s,n(customtrunk),Set(pre_num=${CUT(OUT_${DIAL_TRUNK},$,1)})
exten => s,n,Set(the_num=${CUT(OUT_${DIAL_TRUNK},$,2)})
exten => s,n,Set(post_num=${CUT(OUT_${DIAL_TRUNK},$,3)})
exten => s,n,GotoIf($["${the_num}" = "OUTNUM"]?outnum:skipoutnum)
exten => s,n(outnum),Set(the_num=${OUTNUM})
exten => s,n(skipoutnum),Dial(${pre_num:4}${the_num}${post_
num},300,${DIAL_TRUNK_OPTIONS})
exten => s,n,Goto(s-${DIALSTATUS},1)
exten => s,n(chanfull),Noop(max channels used up)
```

```
exten => s-BUSY,1,Noop(Dial failed due to trunk reporting BUSY -
giving up)
exten => s-BUSY,n,Playtones(busy)
exten => s-BUSY,n,Busy(20)
exten => s-NOANSWER,1,Noop(Dial failed due to trunk reporting NOANSWER
- giving up)
exten => s-NOANSWER,n,Playtones(congestion)
exten => s-NOANSWER,n,Congestion(20)
exten => s-CANCEL,1,Noop(Dial failed due to trunk reporting CANCEL -
giving up)
exten => s-CANCEL,n,Playtones(congestion)
exten => s-CANCEL,n,Congestion(20)
exten => _s-.,1,GotoIf($["x${OUTFAIL_${ARG1}}" = "x"]?noreport)
exten => _s-.,n,AGI(${OUTFAIL_${ARG1}})
exten => _s-.,n(noreport),Noop(TRUNK Dial failed due to ${DIALSTATUS}
- failing through to other trunks)
exten => disabletrunk,1,Noop(TRUNK: ${OUT_${DIAL_TRUNK}} DISABLED -
falling through to next trunk)
exten => bypass,1,Noop(TRUNK: ${OUT_${DIAL_TRUNK}} BYPASSING because
dialout-trunk-predial-hook)
exten => h,1,Macro(hangupcall,)
; end of [macro-dialout-trunk]
```

As you can see, there is a lot of code here just to make a simple trunk call. However, for a coding novice, this will implement all the checks and error handling that you are ever likely to need.

It is impossible to say how much of a performance hit a GUI-based system will incur over a "vanilla" system, as all systems and requirements are unique. However, when planning your system, the extra resources required for a GUI, however small, should be factored into the equation.

Stability

Linux is inherently a very stable operating system, and as a result, many systems with a Linux base, Asterisk included, achieve levels of stability and reliability that would require an awful lot of work and in depth knowledge to achieve in a Windows-based system. However, GUI-frontend Asterisk systems do tend to require more scheduled downtime than non-GUI systems for the simple reason that there is more code running on the server. So whether it is for a patch of Apache, PHP, or MySQL, or a bug is found in the code of one of the FreePBX modules that you have implemented, you are likely to find that over time, a GUI-based system could spend more time offline that a non-GUI one. Although, it is worth emphasizing that in a well set up GUI-frontend, system downtime should still only be measured as a small number of minutes per month, or even per year.

It should also be pointed out that some GUI-frontends, including FreePBX, which we will look at in more detail shortly, separate the maintenance of the GUI from the maintenance of Asterisk, meaning that such updates rarely require a restart of Asterisk.

In addition to planned downtime, there is also the risk of unplanned outages to consider. While a GUI-frontend Asterisk system may be the model of consistency compared to other applications, ultimately it consists of a "vanilla" Asterisk system with some extra applications in the mix. Adding extra applications to any system will usually increase the risk that something will eventually go wrong somewhere.

In order to reduce that risk to acceptable levels, there is an increased need to ensure that appropriate and robust change control processes are in place. These will require that adequate testing of the system, in a separate test environment, is carried out every time a significant change is made. Of course, this should happen regardless of whether or not a GUI is used for administrative purposes, but it's likely that more testing will be required with a GUI-frontend system simply because there are more applications in the mix. This extra testing will cost more over time, but could be mitigated by the fact that the standard code produced by the GUI is likely to reduce the initial development time of dialplan changes. In the end, either your own testing or some independent testing of system stability should ideally be used to inform your discussions of this subject with potential customers.

Restricted functions

It may seem paradoxical to have restricted functions as a downside of GUI-based systems, but that is not the case. Most GUIs, as previously described, enable the use of pretty sophisticated features within Asterisk. For instance, FreePBX contains modules that allow the Asterisk administrator to easily enable call groups, call queues, conference call functionality, and many other features. However, the framework provided by FreePBX and other GUIs can tie you into their way of implementing a particular function, and make it difficult to insert custom code to implement your own functions. Typically, custom code has to be entered into predefined .conf files or into the predefined areas of certain .conf files, or else there is the risk of it being overwritten by the next application of a change in the GUI. So, occasionally you will find that inserting some highly-tuned code is impossible, or at least it has to be done in a far from ideal fashion.

FreePBX

FreePBX is a GUI for Asterisk, the development of which has been led by Philippe Lindheimer. It is far from being the only GUI and may not even be the best GUI, but that it is not a commercial product makes it quite unusual in the range of Asterisk GUIs. It is not a commercial product in this range. Many of the other products in this field are not covered by the GPL and have been developed, among other reasons, to allow businesses to charge for an Asterisk-based system. ScopServ is a good example of a commercial Asterisk GUI. Other commercial products have an open source version that provides restricted functionality and an upgrade path to the paid-for version. Druid from Voiceroute is a good example of this. There is absolutely no problem with commercial approaches, as they should give the purchaser some confidence in the continued development and stability of the GUI, and by association of the whole system.

However, this book has Asterisk as its focus, so it seems logical to consider other open source solutions in conjunction with Asterisk where they exist as a viable option. In the field of open source GUI-frontends for Asterisk, FreePBX stands proud, particularly now that Digium themselves have committed to using FreePBX within their AsteriskNOW offering from release 1.5 onwards.

How it works

FreePBX presents a graphical interface to the system administrators that allows them to configure a wide range of functionality. The functions are separated into modules, each of which represents a distinct function, and most of which can be enabled or disabled so that only the desired functionality is active at any time.

FreePBX does not directly represent the contents of the configuration files on screen. Rather, it maintains its own database in MySQL or sqlite, which is then used to generate certain configuration files afresh every time changes are "applied" to it. This scenario is enabled through the use of `#include` statements. For instance, when changes are applied in FreePBX, the `sip_additional.conf` file will be overwritten, but the `sip_custom.conf` file will remain as is. Both `sip_additional.conf` and `sip_custom.conf` are referenced from `sip.conf` by means of a `#include` statement. In this way, any custom code written to `sip_custom.conf` will remain without the danger that changes in the GUI will overwrite them.

Installation

Typically, FreePBX will come as part of a complete Asterisk package, such as **AsteriskNOW** from Digium, **Elastix** from PaloSanto Solutions, or the freely-available **PBX in a Flash** from Nerd Vittels, and others. There is a very strong argument for implementing an Asterisk GUI in this way—someone else will have gone through the pain of choosing the best Linux distro, choosing the right flavors of Apache, MySQL, and PHP, enabled the relevant options within all those packages, and then installed FreePBX on top. However, it is worth looking at the steps required to install FreePBX should you decide to implement it on a "vanilla" Asterisk system, if only to gain a better understanding of how much extra code is needed to implement it.

At the time of writing, FreePBX development was carried out in the main on CentOS 5.1, so that is their recommendation for a base operating system. Full instructions for the installation of Apache, MySQL, PHP, Asterisk, and FreePBX from this base can be found at the following URL:

```
http://www.freepbx.org/support/documentation/installation/install-
process-for-centos-5-1
```

During this installation, the following packages are installed:

- CentOS 5.1
  - DNS
  - Apache 2.2 (HTTP)
  - Sendmail (SMTP)
  - MySQL 5
- Asterisk 1.4
  - Zaptel 1.4
  - Libpri 1.4
- FreePBX 2.5

FreePBX will run on a number of other Linux distributions too, many of which have documented installation processes at the FreePBX web site. You may also find that, over time, the included packages change to keep pace with developments. For instance, at the time of writing, Asterisk 1.6 has reached the release candidate stage and will undoubtedly form the core of a FreePBX installation once it is finally released.

Configuration

FreePBX requires that a number of basic steps be followed to achieve the minimum working system. They are the addition of:

- An extension
- An inbound route
- An outbound route
- A trunk

Extensions define endpoints, normally handsets, but potentially fax machines or other devices. While not their sole purpose, outbound routes are often used to implement **Least Cost Routing (LCR)**, as they identify which route a call should take based on certain criteria, such as the dialed number. On the other hand, inbound routes tell the system what to do with incoming calls, again typically using call metadata such as the DID or CID. Trunks define the interface with the outside world, and for outbound calls, they may be referenced by one or more outbound routes.

These definitions will allow a telephone on the system to make and receive calls. Any of the other options will implement more complex functionality, but are not essential. However, it is a simple implementation indeed which will require just the options described previously. Let's look at these four definitions in a bit more detail to see how FreePBX implements them.

Extensions

An extension normally defines an internal telephone, either hardware or software, although it can also define other endpoints such as fax machines. The protocol used for extensions is usually **SIP**, although **IAX2** is also used, and Cisco devices have historically had their own proprietary protocol (**SCCP**, colloquially known as Skinny). However, since SIP is by far the most popular one, let's have a look at how an SIP extension is defined in FreePBX.

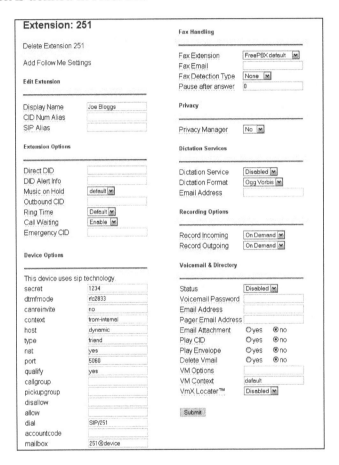

There are a lot of options on this screen, and the use of drop-down lists and radio buttons is made to ensure accuracy.

FreePBX assumes that voicemail will be a desirable function in any Asterisk/FreePBX system. Therefore, it installs voicemail by default, and includes voicemail functions within the extension definition. Simply check the appropriate box and voicemail is enabled with the minimum of fuss. However, if you wish your system to do without voicemail, then the appropriate module can easily be removed.

One impact of defining an extension in FreePBX is that the appropriate dialplan entries for routing calls between internal extensions are automatically added. Therefore, your only routing concern is with calls originating or terminating outside your system, as we will now explore.

Inbound routes

Within FreePBX, the part of the dialplan that handles incoming calls is defined by means of one or more inbound route records. Essentially, this record tells Asterisk what to do with an incoming call. It provides the link between the trunk and the extension. The record can define the actions for a particular DID, a particular CID, or a combination of both. Pattern matching can also be used for DIDs or CIDs so that each possible number doesn't need a separate record. This can be a good way of, for instance, directing calls from a particular region to a local office. Finally, there is a facility to have a "catch-all" record that determines what to do with any inbound call. Let's have a look at an inbound route record, which will transfer all calls through to extension **251**.

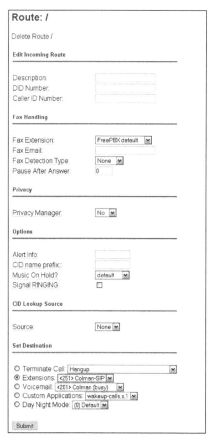

Instead of transferring every inbound call to extension **251**, we can also use other appropriate destinations, but only destinations that have already been defined will appear on the list. In other words, you cannot choose to route calls through to a voice menu, for instance, if you have not already set up that facility. You may notice that there is a **Day Night Mode** option in the example. This is used to toggle between alternative destinations depending on the time of day. You might decide that the calls should go to a receptionist during the day, and to a voice menu system outside normal working hours. Or there may be two separate voice menus—one to handle calls during working hours and one for calls outside working hours. The switch from "day" mode to "night" mode can be time-based, or manual.

Outbound routes

Outbound routes, as their name suggests, define what to do with outgoing calls. If the goal of your PBX installation is to reduce call costs through the implementation of LCR rules, then this is where those rules will be defined. In its simplest form, an outbound route record will determine what happens to calls as they traverse from extensions to trunks. It can be defined for all extensions or a subset based either on an extension number or, more typically, based on the profile of the number being dialed. In our very basic system, an outbound route will look like this:

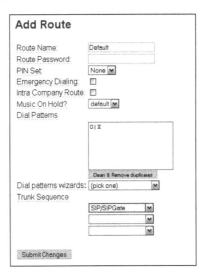

Outbound routes have dial patterns, which are used to match and manipulate numbers. Typically in an outbound route, the dial pattern(s) will be used to identify groups of numbers that you wish to send to a particular trunk. Maybe, for instance, you would set up one dial pattern to match all mobile telephone numbers (which start with "07" in the UK), and then direct them to the trunk you've set up for your GSM gateway.

You may also have noticed that there is room for multiple trunks for a single route. This allows for failover if the primary trunk is unavailable. This is particularly useful if you wish to use a service such as ENUM or DUNDi to attempt to locate a route to a destination before using another route that will incur a cost.

If you wish to start implementing LCR, you will need multiple outbound route records. While relative priority is a consideration with these if more than one route applies to a particular call profile, normally the choice of route is determined by the dialing rules. Therefore, while one route would send all mobile calls via a GSM gateway, another may send all international calls to Asia via an ITSP that offers particularly good rates to that market. However, both routes might have an ITSP with good overall rates as a fall back just in case the preferred option is unavailable.

Trunks

Trunks define the interface to the outside world, and can include DAHDI (Zaptel), SIP, IAX2, ENUM, and DUNDi protocols natively, and other protocols such as H323 or mISDN as "custom" trunks. The trunk definition describes the contents of the relevant .conf file with a very obvious relationship between the options in FreePBX and the possible parameters in the .conf file. It's typical for a very basic installation to have a SIP trunk to a VoIP provider, so let's see how that might look in FreePBX.

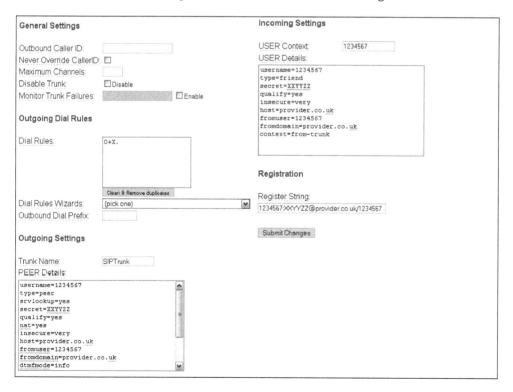

As you can see, there are a few options to consider. While in general, efforts have been made to move away from offering the user a text box where parameters have to be typed, the nature of trunk definitions makes this a little more difficult, so text boxes have been retained for the outgoing and incoming settings. From here they are written to `sip_additional.conf`, which is included in the `sip.conf` file by means of a `#include` statement. The `sip_additional.conf` file is overwritten by FreePBX.

The interesting part of this definition is the **Dial Rules** section. Here, numbers can be manipulated before a call is placed. For instance, in the preceding example, the only dial rule is `0+X`. The `X` matches any digit, so it essentially means "use this rule for any number not previously matched". The `0+` then adds a 0 to the beginning of the number, and the period at the end is short-hand for saying "leave the rest of the number as is". Therefore, the rule as a whole says "add a 0 to the start and place the call". It is also usual to add any calling card prefixes for specific trunks within the dial rules.

You may wonder why we are stripping a zero out of the number early on in the process if it is going to be added back later. However, different service providers can require numbers to be presented in different formats. One may wish the full international format with no leading zeroes (such as 442075555555 for a number in London, UK), whereas another may accept numbers without the international code for UK numbers, but with a leading zero (for example, 02085555555). If you use multiple providers with different requirements, any process prior to the trunk can get very complex unless you strip everything down to a standard format, and just add the provider-specific digits in the trunk definition.

Other records

This is not the place to explore in detail all the options available to you within FreePBX. There are online resources that cover the detail of what each FreePBX record does, such as the FreePBX web site at `http://www.freepbx.org`, and there is also a forthcoming Packt publication on the subject. It would suffice to say that once installed, a system utilizing FreePBX can be very quickly configured to the point where basic functionality is possible. Once that point has been reached, slightly more advanced functions can be easily enabled, such as **IVR** (digital receptionist), **Queues** or hunt groups, **Follow Me**, **Day/Night Control**, and many more.

Summary

Strong opinions abound on the advantages and disadvantages of graphical user interfaces for Asterisk systems. Those who, with some justification, believe that a phone system should sit in a corner and "just work", have concerns over the increased risk of system downtime and the impact on system performance. However, such views tend to be prevalent among those who have come from a traditional telephony background, or are "old school" Linux devotees. The problem is, the world of telephony no longer sits proudly on its own, but rather is converging rapidly with the world of computing, enabling voice traffic to be carried over computer networks, and telephone systems to interact with the desktop. In this brave new world, having the ability to interact with your telephone system from your PC is increasingly desirable.

While adding complexity to your Asterisk system carries a corresponding increase in the risk of unscheduled downtime, the extent of that risk depends on the processes that you put in place to deal with the downtime. In particular, a robust change control environment is needed to ensure system stability, regardless of whether or not a GUI is fronting your Asterisk system. As Asterisk sits on standard PC hardware, standard failover and clustering techniques can also be applied to mitigate the risk. In spite of taking this into consideration, the budget for a traditional PBX or a commercial VoIP system can still be hugely undercut, even if one is not pushing the Asterisk box to its limit.

Ultimately, Asterisk is a hugely-versatile telephony engine. In certain circumstances, such as large installations or high-volume environments, it makes a lot of sense to keep the installation as clean as possible. For such customers, a managed service is quite often the very solution they prefer rather than having to train internal staff to a high level. For other customers, such as the SMEs who make up the majority of businesses, having the ability to carry out day-to-day configuration changes themselves while using a third party for more complex tasks is quite usual. For these customers, who will typically have lower call volumes and require fewer system resources, the performance impact of a GUI is normally acceptable given that it simplifies the administration process.

There is a myriad of options if you decide you wish to implement an Asterisk system with a GUI, although the majority of system configuration and administration GUIs are commercial products. In the open source field, FreePBX is a successful product that is widely used in many Asterisk "distributions", including the latest version of AsteriskNOW from Digium themselves. Like most GUIs, FreePBX does introduce some intricate code into the dialplan, which has the potential to reduce the performance of the server. However, it does allow for a relatively complex Asterisk setup to be created quickly and easily, and is modular in nature so that only the essential features are implemented.

A
Selling Your Solution

Your background prior to becoming an Asterisk consultant may or may not have been technical. Given the nature of the Asterisk software, particularly the fact that it can be freely downloaded, installed, and used with no license charges, it's not that unusual for technical people to progress from "having a tinker" to a position where they have a coherent system that they wish to use as the basis for a commercial offering (of course, remembering that while you are allowed to charge for Asterisk, 'there isn't much point as it can easily be downloaded for free. It's a bit like charging someone for tap water while standing next to a public faucet). Reading this book and absorbing the information and suggestions it contains, coupled with your existing knowledge and expertise, will give you the tools to create an Asterisk system that is a viable commercial proposition—one that can fit into most organizations, large or small. It is certainly the case that there are a great many Asterisk-based systems in wide-ranging environments today, from one-man bands to university campuses. However, creating a system capable of doing a great job for your potential customer is only half the battle, you also have to sell it to them.

In this appendix, we will look at some of the challenges you may face as an Asterisk consultant attempting to make sales to small, medium, and large enterprises. The focus will be on selling to **small and medium enterprises (SMEs)** as sales to large organizations would normally require the vendor to demonstrate significant turnover, stability, market experience, and support capabilities. It seems fair to assume that a vendor with those capabilities will already have a dedicated sales function. However, many of the principles mentioned will apply across the board.

In the beginning ...

Before you even think of hitting the market, there are a few questions you need to ask yourself. Top of that list is the most important one—Do you truly "believe" in your product? In other words, are you absolutely certain that when you install your system, not only will it do everything the customer's previous system did, but also actually offer them extra functionality and a commercial advantage over competitors who do not have your telephony system? If you do not have this belief, then you need to understand the fine detail of what your system can do in order to give it to you, because if you go to a potential customer and the questions start getting tough, only this absolute belief based on demonstrable facts will see you through.

However, you have just learned how to construct a high-quality, extremely-stable telephony platform with excellent functionality, and through its open source nature, the ability to seriously undercut the price of commercial offerings and still make you a profit, so why wouldn't you be confident?

So now you're fired up about Asterisk and its capabilities and possibilities, but how do you go about selling it to potential customers? Again, before you hit the market, you need to accept that in order to be successful, not only do you have to provide one or more Asterisk servers and a bunch of handsets, but you also need to ensure that the ancillary aspects to any telephony system meet certain required standards if the installation must achieve the customer's goals. Items such as the LAN, WAN, and Internet connectivity need to be of a certain standard, possibly using VLANs and/or QoS to ensure voice quality. The environment the server will reside in should be secure, clean, and air-conditioned, with power failover in many cases. The customer may wish to implement some or all of these aspects themselves, but you will need a mechanism for determining the minimum standard of each, and of allocating responsibility fairly once the system and maintenance agreement is in place. You should also be prepared to provide any or all of the prerequisites if you are asked to do so, which may mean having agreements in place with companies who can do a job to the required standard, if you do not wish to do it yourself.

Drivers for changing phone systems

Now you're ready to hit the market, so let's have a look at why potential customers may consider a new telephone system. For 99% of the time that you speak to a customer, the main reason that they are considering a change is that they've heard VoIP will save them money on their calls. However, a little probing will often reveal that this is not the sole reason. Looking at the SME market primarily, you may well find secondary drivers such as:

- Issues with the current telco's service: These include issues such as taking hours or even days to sort out any problem no matter how small.

- Cost of adding new phones: With traditional twisted pair telephony, adding a new phone or moving an existing one can be expensive. While people have become used to running CAT5 cable, twisted pair cables seems like black magic still.

- Number portability: The customer is moving to a new location and wants to retain their number.

- Maintenance costs: A lot of telephone companies derive a huge amount of their income from maintenance fees. These can be considerable and are often overlooked by the customer.

- Flexibility: Most traditional systems "can" be quite flexible, but often a specialist engineer is required to access the system remotely, or even physically, and enter obscure codes, simply to change something as simple as ring patterns. The charges made for this often deter the customer from requesting changes that would benefit their business.

These drivers cover most of the reasons that most customers consider changing their phone system. However, there may well be other reasons a change is being considered — for instance, an office move often prompts consideration of telephony requirements. The relative priority of the drivers will vary depending on other factors, an obvious example being the financial climate at that point in time. If you want to be successful in targeting your product accurately, it is important that you understand which drivers are most important to your target market when you are contacting them.

A word on cost

If you were to have a discussion with some people about the USP (Unique Selling Point) of modern IP-capable PBX's, the majority are likely to mention that it is the ability to make cheap calls. But, often this is not the case. It is becoming more usual for telcos to offer bundles or even flat-rate pricing for PSTN calls, along the lines of the charging strategies in the mobile/cell market. As a result the cost difference between PSTN and VoIP calls is diminishing, sometimes to the point where huge volumes of calls are needed to make a compelling case for change when only call costs are considered.

However, in the SME market, significant cost savings can be made simply by reducing the number of PSTN lines on which a company needs to pay rental. Remember, line rental is paid whether or not any calls are made over the line. In addition, it's not unusual for customers to be locked into minimum contract terms, some as long as seven years! This may save some money on installation charges, but that saving is completely outweighed by the cost of line rental over that period, particularly if one or two years into the contract a means of reducing the number of lines required is found. If you have a potential customer in that situation, they will need to have a frank discussion with their line provider about early exit from the contract, and the cost of that will need to be factored into the ROI calculations.

To illustrate the potential line rental savings, consider a hypothetical dispatch company. They have five lines so that they can deal with multiple calls at certain times of the day. These lines are really busy from 7 am to 9 am while they deal with deliveries, and busy again from 3 pm to 5 pm while they deal with pickups. The rest of the time, there's probably hardly a call, but they're paying for the five lines 24/7. If you can approach that company and say, "Guess what, you can do away with four lines *permanently*, do you think they're going to be interested?" Of course they are, because by eliminating line rental, you're disposing of a significant fixed cost.

So you can see that, when selling the potential to save money to a prospective customer, it's not as simple as telling them that they'll save a certain amount on all their calls. Depending on the company's call volumes, it's possible that the ongoing line rental savings will be the major cost benefit, and to quantify that you will need to know how many lines they have now and how those lines are used. You can then put together a proposal illustrating how many lines the new system will need, with associated reduction in line rental charges, and how their call profile will be represented using VoIP where appropriate. If this also results in significant cost savings, then it's even better.

A word of caution, though—all of the major telcos are reducing their call charges. It's becoming a downward spiral. You, or your chosen ITSP, cannot hope to compete on a cost per minute with a telco with millions in the war chest. Indeed, it is the current consensus that call charges will disappear completely in the next few years to be replaced by bundled minutes. It would not be wise to make long-term call cost savings the cornerstone of any proposal you put forward.

Generating interest

So now you have your core product, and you have the confidence that it can bring significant benefits to customers. The next step is to make people aware of that, to get this message out to potential customers. Make no mistake, though, marketing your shiny new Asterisk system is hard. This is regardless of how good it is, how cheap it is as compared to competitors, how easy to install and maintain it is, and how wonderful the benefits to the customer are. It's hard because you have to get that message out to the right people, at the right time. Unfortunately, there's no sure-fire way to make it work either. However, we have a couple of pointers that will hopefully get you started on the right lines.

With any telephony system, including Asterisk, the important thing to recognize is that they are not changed very often. We've seen already some of the common drivers for changing a phone system, but it's also worth having a look at a couple of other factors considered by companies making that decision.

- Change: Changing a phone system has the potential to cause huge disruption to a busy office environment. Staff will probably need to gain familiarity with new handsets, new processes (for example, for setting up conference calls), new functionality, and maybe even new extension numbers.

- Risk: Phone systems continue to be a core part of every company's communications strategy, and the disruption of that service even for a short amount of time has the potential to cause serious loss of income and goodwill.

As a result of these factors, messages along the lines of "Simply install this new phone system to realize savings in line rental charges" are unlikely to work unless you happen to catch a decision maker at just the right time. In order to persuade a company to even consider changing their phone system, you need to be able to demonstrate that the change is worthwhile and will be completely stress free. Customer testimonials are great here, but when starting out, you may need to take a different tack, such as demonstrating that you use a formal methodology for managing the implementation, such as PRINCE2 or PMP.

But to start with, you need a flow of hot leads into the business. It's very sensible not to have all your marketing eggs in one basket. A strategy that combines alliances, advertising, and more targeted contact over a range of media is more likely to bear fruit than a single mailshot to all the businesses in a particular area. So how do you go about it?

Alliances

Forging an alliance (formal or otherwise) with another company is a strategy for generating leads that, with a little luck, can be very successful. For instance, you may already have an arrangement with a cabling company, through which you can utilize their services if customers need their LAN upgraded. However, in the normal course of their business elsewhere (for example, outfitting new or refurbished premises), they may have customers who would welcome the details of a company that can provide an affordable, high-quality phone system.

An even better association would be with a commercial property agency, as they are involved with businesses at precisely the time that most consider their telephony options. The benefit for them may merely be the chance to offer a better "package" to their customers, or possibly you would need to negotiate a finder's fee for every customer they bring to you.

Advertising

Advertising is about getting your "name" out into the marketplace, preferably as a provider of high quality yet affordable systems. Traditionally, advertising has been seen in terms of television, radio, newspaper, and magazine ads. While adverts in select trade magazines can be an effective strategy for marketing an Asterisk system, some online tools are likely to be of more use, and it is into this area where much of the marketing effort goes these days. The great thing for you is that online marketing can be carried out without huge cost, and yet achieves far better results than "old school" methods.

Search engines

A good exercise when designing any part of your marketing strategy is to put yourself in a potential customer's position. For whatever reason, they have made the decision that they want to consider changing their phone system. So what will they do to investigate their options? To start with, they are likely to use one or more of the popular search engines. Therefore, having your details come up on the first page of results, or in the first lot of paid links, is likely to generate a steady stream of very hot prospects.

However, taking a step backwards, in order to be included on search engines, you need to have a presence on the Web. Having a web site does not have to be costly, and if done well, it can project exactly the professional image that is required. Although, in order to feature highly in search engine results, your web site not only has to have the right content, but must also be designed in such a way that it will feature highly in search results. Additionally, it is hugely beneficial if your web site is linked-to from many other places. There are a number of ways of achieving this, one being the use of blogging as a business tool.

Advertising on search engines is slightly different, in the sense that the emphasis is very much on choosing appropriate keywords to associate with your advertisement. The skill here is in choosing keywords that are likely to be used by prospective customers, but haven't been chosen by rival phone system providers for their adverts. This increases the likelihood that your advertisement will appear more often when one or more of your keywords are entered. The pricing structure for such advertising is based on the number of times your advert is clicked, hence the term **Pay Per Click (PPC)**.

Putting everything together into a coherent whole requires a good deal of expertise, so this is one area where using the skills of a company that specializes in **Search Engine Optimization (SEO)** is usually money well spent.

You can go to the following web sites for further reading:

www.searchengineguide.com

www.searchenginewatch.com

www.sitepoint.com/kits/sem2/

Become an expert

Another means of raising your profile in the marketplace is to position yourself as an expert in the field of Asterisk-based systems. After all, if you have a product that is commercially viable, then you've progressed to a point where you have knowledge that is relatively rare. One means of publicizing that expertise is to write, possibly in the form of magazine articles, press releases, or even a book. Alternatively, you could offer tidbits of information via your blog, which has the added benefit of bringing people to your web site and generating external links. You could also use this strategy to initiate an email marketing strategy, through which you ask web site visitors to subscribe to a regular newsletter.

Relationship marketing

By acquiring information about your audience, you have reached a stage when you are no longer using a scatter-gun approach to marketing, that is putting a message out to a large audience in the hope that some of them could use your product. Now you can target your message based on the information you have about individual subscribers. At this point, tradition suggests that you start reeling in your potential customers, and try to convince them to change their phone system. **Please resist this urge**! By all means, inform your captive audience of the benefits of Asterisk in general and your wonderful system in particular, but this strategy works best if you are perceived as an expert in your field, and not as a sneaky sales person (apologies to all sneaky sales people reading this!). In other words, they need to trust you.

Email as a marketing tool

Now, the fact that your prospective customers are interested in your opinions indicates that it's likely they are considering a new system anyway. So let that itch continue to irritate, and when they finally decide to scratch it, they will come to you for advice, precisely because you've worked hard to build up that trust. Of course, since you are communicating directly with them through the use of personalized emails (at least that's the impression you are trying to give them), there is an opportunity to foster that relationship. An effective email marketing campaign can track what each recipient does with the email you send them. Do they open it at all, how many times do they open it, and which links do they click on? The last in particular is very pertinent, as it can allow you to understand where your subscriber's interest is in a better way. Do they click through to a list of handsets? Are they looking at how to improve their LAN? All this is useful data that can help you build up a picture of subscribers and tailor the message you send them.

You may be worried that doing all this writing rather than actively selling to customers will simply take up too much time. It is certainly likely that you will spend a few evenings updating your blog, composing marketing emails, and understanding the results of previous campaigns. However, selling is not possible if you are not generating leads in the first place, and it is far better to be feeding hot leads into the sales funnel rather than a load of dross. In addition, the research needed to generate content for your blog/emails will keep you up-to-date with what's happening in the marketplace, which you should be doing regardless.

 Perry Marshall is a guru of Google advertising and email marketing techniques and shares much of his knowledge on his web site www.perrymarshall.com.

Tracking prospects

In doing all of this high-quality online marketing, it's extremely easy to lose track of where you are with it all. It is highly recommended that you implement a system that will allow you not only to record what has been done (preferably as easily and automatically as possible), but also to drive the marketing effort. Although, the important aspect of whatever system you implement is that it is relatively easy to use and that it doesn't get in the way of getting the message out.

Many people starting out in business use Outlook in the first instance, as it allows them to record prospect details, and, with the addition of Business Contact Manager, run email campaigns and track some results. However, Outlook with Business Contact Manager is not a fully-fledged CRM solution, so we would recommend looking at an alternative. Throughout the book, we have championed the use of open source software, and in that arena, one of the best CRM solutions is SugarCRM. In fact, some Asterisk-based systems have SugarCRM bundled as an optional part of the installation. Of course, there are other open source solutions too, such as vtiger, Compiere, Concursive, and others.

While SugarCRM may take a bit longer to configure for your particular purposes in the first place, it is more likely to cope with a high degree of growth over time, saving you the disruption of switching from one system to another. Within SugarCRM, you can define marketing campaigns over varied media with or without prospect lists. So whether you're running a magazine ad or a highly-targeted email campaign, you can record the information in SugarCRM and track what happens after the campaign runs. For campaigns with prospect lists, you can record relevant activity manually (useful for phone calls or inbound emails). However, it also has the capability to run an email campaign for you, from sending personalized emails to tracking all activity, including clickthroughs. Of course, all this activity is recorded automatically, and available in future if the prospect calls up. This can really add an element of professionalism to even the smallest Asterisk consultancy.

For further reading, see *Implementing SugarCRM,* Michael J.R. Whitehead, Packt Publishing.

Of course, once a prospect becomes a customer, SugarCRM has all the functionality needed to track activities with that customer, including support tickets.

Converting the prospect into a sale

So, you have implemented an effective marketing strategy that feeds hot prospects into your lap. They have a need for a new phone system and think that you may be able to provide it. You've arranged to meet them, but how do you go about converting that hot prospect into a customer more often than not? The best marketing in the world will be undone if you make a bad impression in person.

Determining your customer's hardware requirements

When you go to your customer's site, your desire might be to impart knowledge, but in a sales situation it's far more important that you listen. Remember, you've already built up a degree of trust through your marketing efforts, and as a result, the customer has brought you in to use your expertise to solve a problem, "not" to be a sales person. Leave that to the cold callers whose conversion rate is probably in the region of one lukewarm lead per 100 calls, requiring 5 to 10 site visits before you get a sniff of a sale. Whereas, when you are asked to share your expertise as a consultant, you will have the hottest lead possible and your conversion rate will be nearer 1 in 2! So your main goal is to determine the customer's needs and wants by listening attentively to what they are saying.

Of course, during your visit there is certain information that you will need to get if you are to provide an accurate proposal, but you will usually find that the majority of that information comes out naturally anyway, with little or no prompting. However, designing a requirements form makes a lot of sense and adds to your professional image. In addition, it means you have all the information you need written down, and are not relying on your memory once you get back to the office. On your form, you need to discover certain items, including:

- How many lines do the customers have, and more importantly how many do they need?
- How many extensions?
- How many calls are internal and how many external?
- Do they need voice recording?
- Do they need CDR?
- What are their cabling needs? Is CAT5 already there?
- Do you need a switch, does it need to be **PoE(Power over Ethernet)**?
- Where are you going to site the PBX?

The key to providing a proposal is in the first three items. You need to do your homework.

- Is their existing Internet access sufficient?
- If not, what is required to carry their potential IP call traffic?
- If appropriate, how good is the broadband in their area?
- How powerful a processor will you need?

The number of extensions will affect your costs. Remember you will need to program each phone, and every phone adds to the complexity of the dialplan if you are providing a system without a GUI.

Choosing the right phones

Which phones you use depends on so many factors, such as your customer's budget, your experience with particular makes and models, the intended usage, ease of provisioning, and so on. We are not going to tell you what phones to use and what phones to avoid—for a start, trashing a manufacturer's phone might land us in a whole heap of legal issues! However, we can say that, in general, you get what you pay for.

Tales of woe

In 2005, we had a chance to run a pilot install consisting of four incoming lines and four extensions. A Digium card was chosen to handle the four analog lines. We selected budget-priced phones which looked to have the features we needed.

Within days, the users were complaining of echo issues. Typically, these kinds of problems lie with the card, so we ran all sorts of diagnostics, playing with the gain, and so on. After much fruitless tweaking and head-scratching, it turned out that the microphone sensitivity on the handsets was way too high and was causing feedback loops. We very nearly lost the contract until we opted for more expensive phones. Interestingly enough, some four years later, there is a current wiki discussing the same issues.

Rather than a lengthy discussion on all of the phones out there, here is a selection of some that have worked well for us. You may like other phones, which is fine, the choice is up to you. Of course, you should make sure you are familiar with the idiosyncrasies of each phone before you add it to your "approved" list.

Aastra

The new 5 series Aastra phones are simply the best we've come across from a price/functionality point of view. In the past, they have had issues staying connected to remote-hosted PBXs, but Aastra claim to have addressed this shortcoming.

Linksys

Linksys SPA-942s are great hosted phones (connecting to a remote PBX), but their lack of BLF (**Busy Lamp Field**) capabilities on Asterisk as well as speed dials, is an issue. As with most phones, the firmware is updated quite regularly, so it is worth keeping an eye out to see if its shortcomings are addressed in the future.

Siemens Gigaset IP DECT phones

A well priced phone that works well as a "walk about" phone (either internal or hosted). Earlier systems, such as the Siemens C460 IP, didn't like to handle more than one SIP provider or one handset per base station. However, most of the current range will cater for six handsets per base station. The station itself will handle three simultaneous calls (two VoIP and one fixed).

Snom M3

Also suited to small office requirements is the Snom M3, a relatively new entry to the business of DECT market. This phone comprises of a base station and up to eight handsets. The handset, although quite small, has good voice quality and a usable speakerphone. Each base station can handle up to three simultaneous calls, so this is not a suitable solution for an office with high call volumes. The base station has a good range, although the claimed 50 m indoors/300 m outdoors is likely to be slightly fanciful. However, it is possible to add DECT repeaters into the mix to extend the range if it is inadequate. Each handset is tied into a single base station, meaning that roaming between base stations is not possible. Such functionality tends to be the preserve of significantly more expensive systems.

Remote support

One area of high cost in the traditional PBX market is the need for providers to employ lots of engineers to go out and fix, or reconfigure phone systems for the customer. If you are pitching to replace one of these ancient systems, a significant part of your proposal should be around the remote monitoring and maintenance services you can offer, as this is relatively easy to achieve with an Asterisk-based system. As virtually any support or maintenance task not involving the need to physically move equipment or components can be carried out remotely, highly-responsive SLAs should be possible.

Make it secure

When specifying the remote support channel for customers, make sure you can secure the access with a tunnel, whether it's SSH or some other method. Remember, you might also need web access to reconfigure individual phones.

Do's and don'ts

Over the years, we have accumulated some tidbits of good and bad practice when selling Asterisk-based systems. They are presented here for your consideration, but, as always, you should only use what is applicable to your situation and needs.

The do's

The practices you should use while selling Asterisk-based systems are as follows.

First impressions

It goes without saying that you need to look smart (but not too smart as you are a consultant and not a sales person, remember!). Got an old car? Maybe you should consider parking it away from the customer and walk. Get there early, you can hang around the reception for 15 minutes and gather your thoughts.

Get brochures printed

It doesn't matter if you wowed the customer at the meeting, they're going to take time to think over your proposal. Nothing beats a nice brochure to look over later, which they can pass to colleagues.

 Get some good quality folders printed that take inserts. This way, if you get new products, you only need to reprint the inserts and not the whole pack. The copy for the inserts can also be used on your web site.

Take notes

Design a questionnaire sheet that you can fill out during the meeting. Customers like to think you're listening to them. It will also form the basis of your quote.

Send the quote in a timely manner

The customers like to think that their potential business is important to you. Make a habit of doing the quote as soon as you get back to the office while the information is still fresh. Have a standard email detailing the benefits, but unless it's really urgent, don't send it the same day unless requested. (Don't want to sound too desperate, do we?)

Follow up the quote

You might think that you've done such a superb pitch that the customer won't go anywhere else, but guess what... they're fickle. If your competition walks in a few days later, you may very well be forgotten. So you should call the customer three days later to confirm they got the quote and ask them if they have any more questions. Agree to call them back in a week if they're still not sure. Use your CRM system to put the calls in your diary and ensure you make them. That one follow-up call can make the difference between closing the deal and losing it.

Target the decision makers, but don't ignore IT

Your initial contact at a company may be the IT department, which is fine, as at the very least it gives you the opportunity to alleviate any technical concerns up front and gives you the opportunity to gain their "buy in" to your system as a great solution. But you quickly need to establish if they have the authority to make a decision, and if not, who does. In smaller companies, this will probably be either the Managing Director or the Finance Director. It is to them that you will sell the operational benefits, and/or the cost savings.

If IT is not involved from the start, then you need to talk to them as early as possible in the process, as they will be heavily involved in the implementation of the solution and probably the ongoing management of it too.

The don'ts

Of course, there are some pitfalls that you should try to avoid too.

You don't need a fancy office

Asterisk-based systems are frequently installed, in part, to allow companies to more easily enable remote working practices. In other words, one of their switchboard operators could very easily be fielding calls from home. If you are starting out as an Asterisk consultant, wouldn't you use that same principle to allow you to work from a modest rented office? For a small monthly, fee you can rent one or more hosted servers that can run your PBX, web site, and CRM so that you can access them wherever you have Internet access.

Most of the time you will travel to customer sites, but if you need to host a meeting, there are many places where you can rent a meeting room by the hour.

You may not wish to divulge this setup to a prospective customer on your first visit—old prejudices can die just as hard as old habits. However, if further down the line it becomes apparent, then there is no reason to deny it or even apologize for it. After all, you have proven that your system enables remote working in a means that is transparent to the caller.

Don't cut corners on the solution

Customers are happy to get a system at the lowest possible price, but when it stops working and they can't do business, everything goes out the window. A small monthly saving on line rental and call costs pales into insignificance if their phones are down for a day. Make sure you carefully point this out as tactfully as you can.

One example is with Internet circuit bandwidth. You might find that the customers don't want to separate voice and data traffic, as they claim to be light users who never approach their bandwidth limit. If that is the case, all well and good, but you need to make them aware of the consequences of their decision so that there is no comeback. Express your concern, and say "OK, but if the sound quality suffers, you'll need more bandwidth or VLANs, or QoS". Put your recommendations in writing; this way you're covered and the customer can't blame you.

Don't under price

There is an old saying:

> *turnover is vanity, profit is sanity, but cash flow is king!*

The simple fact is that, good as Asterisk is, it still needs some level of TLC. You need to factor in your ongoing support, so don't price yourself too low. To quote another cliché:

> *the most expensive isn't always the best, but the cheapest is nearly often the worst*

If your quote is too low, a customer will think it's too good to be true and you'll be unlikely to win the deal. More often than not, they'll go for the quote in the middle. That's where you want to be. However, you can alleviate some concerns about a low quote by being totally transparent about your pricing. If the customer sees that certain aspects of your system are considerably cheaper, such as the lack of software maintenance charges for instance, then they can be reassured that your installation and support services are not "cheap and nasty". Although, be careful as this requires your competitors to be similarly transparent, which may not be the case.

Don't have a huge margin on handsets

It's worth remembering that your customer can quite easily source handsets themselves, so stay aware of retail prices and resist the urge to boost your margin by overcharging for them. However, it is perfectly acceptable to add a transparent charge for assembly, and, if you are not auto-provisioning, a charge for installation as well, which will apply regardless of whether you supply the handsets.

Don't supply a PC as the phone server

In smaller companies, the traditional PBX has tended to be a plain metal box with a few vents, attached to the wall. Now you and I know that Asterisk runs on standard PC hardware, but it does not mean that you should turn up on a customer site with a PC under your arm to install as their new phone server. To your customer, a PC is something that sits on or under their desk, and every so often, causes them a lot of grief and frustration. You are not going to give them a "warm, fuzzy feeling" by sticking one in their server room (which so happens to double as a broom cupboard).

If you're selling to a range of customers, you need a range of server products. It is perfectly feasible to construct a rock-solid Asterisk box using server-quality components, and put it into a wall-mountable case with your logo on it. If you're smart, the same components will fit in a 1U or 2U rack-mount case too, and into a solid, floor-standing server case. Then you have three product lines where the only difference is the case. You can then introduce more options by changing the innards, so you may have entry-level, mid-range, and large systems aimed at different sizes of customer.

The reality, though, is that all but the smallest customer will require and expect a rack-mount chassis for their PBX. For these customers, it is usually better to utilize existing server lines from well-known providers such as IBM, HP, or Dell. Attempting to construct such a server yourself will not save you much money (if any), and you will not have a manufacturer's warranty for the system as a whole to fall back on.

If you decide you wish to construct your own PBX servers, in the long run, it is worth your while using good quality components, that is server-grade hardware. It is likely to increase significantly the **MTBF (Mean Time Between Failures)** of the PBX as a whole (of course, this will be the same as the shortest MTBF of the components making up the system), and will also tend to give you longer warranties on the components. For instance, a typical consumer motherboard has a 12-month warranty, but the server-grade version of the same motherboard has a five-year warranty! Although, you may be constrained by the availability of server-grade motherboards in the right form-factor for systems aimed at very small businesses. In truth, this is a compromise that you will have to decide upon carefully.

For some installations, a simple wall-mountable case that holds a mini-ITX motherboard is appropriate. It's the size of a ream of A4 paper and, with the addition of a logo, certainly looks the part. Installing something like the VIA CN 1000 1 GHz Fanless (less moving parts, less likely to break) motherboard makes it an ideal Asterisk platform for up to 15 simultaneous calls.

Summary

This chapter introduces some effective marketing and selling techniques for Asterisk-based systems. It's not intended to make you into the hardnosed salesperson, but it hopefully gives you some insight as to what is likely to work with regards to selling your solution. It doesn't matter how technically competent you are, if you can't sell your product, the only way is down.

Modern telephony solutions present an opportunity for organizations to cut costs while still maintaining or improving efficiency. At the time of writing, the world economies are in a tail spin, but remember this—in a downturn, there are always a few who benefit. You could be one of those few, as there has never been a better time for promoting the cost benefits of an Asterisk-based system.

In Appendices B and C, you will find information you might want to include in sample emails when pitching, as well as a sample appointment sheet. These are only a couple of the many pieces of the jigsaw that will need to be in place before you have a highly-effective marketing, sales, and implementation process. Some hard work is needed to put it all in place, but once there, it will keep on generating satisfied customers.

Sample Email Content

What is VoIP?

VoIP stands for **Voice over Internet Protocol**. It's simply a means of sending voice over the Internet via a broadband connection, or an internal network. It uses the same cable standards used to connect your computers. In much the same way as an MP3 player digitizes music, VoIP does the same for voice. Once converted, it's simply data that can be transmitted around the world, literally!

Because it's based on the Internet, VoIP is incredibly reliable, so much so that BT (British Telecom) is in the process of converting all of their exchanges to the same technology. Buying a legacy phone system using landlines today is like buying an analog TV set just before the whole country goes digital. Yes, it will work with various adapters, but you're missing out on many of the new features.

The big monopolistic telephone companies have been around for a hundred years now, and in that time, the way phones are used hasn't changed much. Why should these companies innovate when they can simply sit back and rake in the money? Much the same goes for the traditional phone supplier. They've been quite happy selling line rentals for years and making a margin on every line rental that's installed.

But all that is changing and the phone companies know it. It is not uncommon to hear of seven-year line rental contracts! While these seem to save the customer money, the real reason these packages are being sold is to lock the customer in, so that when that same customer realizes he can have 10+ calls on one broadband line, there's nothing they can do about it. Do you really know where your business will be in two years, never mind seven?

Why should I consider VoIP?

VoIP is considered to be futuristic technology, more so than the mobile revolution was back in the eighties. Why? Because it's set to completely change the way you think about your telephone calls. No longer are you constrained about what your local telephone exchange can do—you can make the phones work the way you run your business, and not make your business run around how the phones work.

It's all about convergence. No longer should you consider your phone and data traffic separate. Now they can interoperate with one another. It is now perfectly possible to prioritize calls coming into a company based on information that the company holds about the caller.

Ultimately, moving to VoIP is not just about cost savings, but having the ability to significantly improve the way you run your business.

Cost savings

The payback time on a VoIP system could be a matter of months depending upon the system employed. With a traditional phone system, there is no payback! How is this achieved?

Call costs

Many destinations around the world are 1p/min and significant cost savings are to be made to mobiles. However, it is our belief that this saving, while great, will be short term. Before long, paying for a phone call will be as alien as paying for email.

Line rental costs

Here, significant ongoing costs savings can be made. Given that even today a single broadband line can support up to eight calls with full PSTN quality, or fifteen calls using the compressed GSM codec, many fixed line rentals become redundant.

Wiring costs

Because VoIP runs on CAT5 network cabling, separate phone cabling is not required. In addition, if a phone is required in a new location, moving it is simply a case of unplugging it and taking it to the new location. You won't need a telecoms engineer to run a new cable and reprogram the phone system.

Reduced infrastructure costs

VoIP is not location specific, therefore remote satellite offices can be part of a central telephone system. In smaller branches, a local phone system is not required, as they simply become a remote extension of the main site. Remote/home workers simply need an IP telephone, and using their broadband connection, they become part of the same system with completely free "internal" calls, even if calls are made to the other side of the world.

Centralized management

If you're reducing costs, it doesn't mean that you have to give anything up. In fact, you gain in many significant ways. Just as you can control your data network from your servers, now you can do the same with your voice network.

High granularity reporting and analysis of system usage is now yours in real-time. This means you can see traffic patterns, abuse, and employee performance at a glance.

System integration

Once the preserve of companies with deep pockets, many VoIP systems run using open standards, which means you can efficiently integrate the system with many business applications and databases. Such uses could include the ability to:

- Pop up customer details before you answer a call
- Click to dial from the desktop
- Record calls
- Store contact lists in company directories accessible from the phone

Unified messaging

This holy grail for the major telcos that was never successfully achieved, is now possible at the company level. Email, voicemail, fax, and presence can now be accessed from one location.

Reliability

In the year 2006, for the first time, VoIP systems outsold traditional phone systems. It is highly unlikely that this would be the case if there were concerns about reliability, and let's face it, BT are spending some eight billion pounds converting their exchanges to IP. Reports of poor quality are invariably down to improperly configured and specified systems.

Closed and open systems

VoIP is relatively new, in the sense that it's been around for approximately ten years. As ever, there were many competing standards, and one primarily open standard — SIP (**Session Initiation Protocol**). The last eighteen months (at the time of writing) have seen many companies moving over to SIP and abandoning their proprietary protocols. As a result, there has been an explosion in the number of high-quality handsets available.

Superior sound

Traditional phones are restricted to a given bandwidth, and apart from greater clarity, this bandwidth hasn't changed for a hundred years. On the other hand, VoIP is able to take advantage of the advances in digital processing and wideband codecs are now available, which are the equivalent of HD voice.

Fallback solutions

With a traditional system, if builders inadvertently dig up the phone lines outside your office, you could be without phones for days. With a VoIP system, your connection can be monitored 24/7 automatically. In the event that communications fail, a prearranged fallback solution may be engaged. This could range from broadcasting all calls to predefined mobile numbers and/or alternative fixed line numbers, or diverting calls to another office. All of this can happen within seconds with no manual intervention.

Broadcasting calls

A unique feature of VoIP is the ability to broadcast calls. This means that any inbound call can be sent to multiple destinations (for example, office extension and mobile/cell phone) at the "same" time.

A number for life

We're all used to keeping our mobile number even if we change mobile operators, but with fixed line numbers, move the location of the office half a mile from where you are now, and in all probability, you would have to get new numbers because your new location happens to be served by a different exchange.

Number porting

Most telephone numbers can now be ported to IP. This means that even if you relocate to another part of the country, you take your numbers with you!

Local numbers

Recent surveys show that most customers prefer to deal with local companies, and some even refuse to answer out of area calls. With 90% of all area codes in the UK now available for VoIP number allocation, you can present a local number even if you're not physically located in that area.

About XYZ

Below is a sample email you might want to draw inspiration from:

Our philosophy

At XYZ Solutions, we specialize in all aspects of IT systems, telephony, and networks—whether it is designing and implementing your company network, or simply maintaining your network for you.

As a one-stop shop for business computer systems, we can address all of the IT needs of your company—from system design and analysis, security, networking, virus protection, disaster recovery, through to supplying hardware and software. We supply and install the Asterisk VoIP PBX system.

Our passion

Our passion is to improve your business performance by providing a combination of sound IT advice gained from 20 years of experience with "hands-on" technological expertise. Our focus is on increasing your productivity and business efficiency. Along with pioneering work with VoIP technology, we can provide you with the additional benefits of enhanced customer service, whether for businesses or consumers, and significantly reduced telephone call charges.

We have a "can do" attitude and take pride in our work, offering the most cost-effective solutions to your business needs, but above all, ensuring that they're right for you and that they actually work.

C

Sample Appointment Sheet

DATE: _____

COMPANY: _____ CONTACT NAME: _____

TELEPHONE NO: _____

EMAIL ADDRESS: _____

Q 1.   What business need is driving this change?

_____

_____

_____

_____

Q 2.   How many phone lines do you currently rent? (phone, fax, broadband, and others — analog or ISDN?)

_____

_____

Q 3.   Who is your current Internet circuit provider?

_____

Q 4.   What is your Internet circuit bandwidth?

_____

Q 5.   How many locations are involved?

_____

Q 6.   How many extensions do you have?

_____

Q 7.   How many extensions per location do you have?

_____

Q 8.   Do you have any free network sockets and is there power at each phone point?

_____

_____

_____

Q 9.   How many home workers do you have?

_____

OTHER:

_____

_____

_____

_____

_____

_____

_____

Index

Symbols

A

Thank you for buying
Asterisk 1.4

Packt Open Source Project Royalties

When we sell a book written on an Open Source project, we pay a royalty directly to that project. Therefore by purchasing Asterisk 1.4, Packt will have given some of the money received to the Asterisk project.

In the long term, we see ourselves and you — customers and readers of our books — as part of the Open Source ecosystem, providing sustainable revenue for the projects we publish on. Our aim at Packt is to establish publishing royalties as an essential part of the service and support a business model that sustains Open Source.

If you're working with an Open Source project that you would like us to publish on, and subsequently pay royalties to, please get in touch with us.

Writing for Packt

We welcome all inquiries from people who are interested in authoring. Book proposals should be sent to author@packtpub.com. If your book idea is still at an early stage and you would like to discuss it first before writing a formal book proposal, contact us; one of our commissioning editors will get in touch with you.

We're not just looking for published authors; if you have strong technical skills but no writing experience, our experienced editors can help you develop a writing career, or simply get some additional reward for your expertise.

About Packt Publishing

Packt, pronounced 'packed', published its first book "Mastering phpMyAdmin for Effective MySQL Management" in April 2004 and subsequently continued to specialize in publishing highly focused books on specific technologies and solutions.

Our books and publications share the experiences of your fellow IT professionals in adapting and customizing today's systems, applications, and frameworks. Our solution-based books give you the knowledge and power to customize the software and technologies you're using to get the job done. Packt books are more specific and less general than the IT books you have seen in the past. Our unique business model allows us to bring you more focused information, giving you more of what you need to know, and less of what you don't.

Packt is a modern, yet unique publishing company, which focuses on producing quality, cutting-edge books for communities of developers, administrators, and newbies alike. For more information, please visit our website: www.PacktPub.com.

Building Telephony
Systems with Asterisk

An easy introduction to using and configuring Asterisk to build feature-rich telephony systems for small and medium businesses

David Gomillion   Barrie Dempster   PACKT

Building Telephony Systems
With Asterisk

ISBN: 978-1-904811-15-2   Paperback: 180 pages

An easy introduction to using and configuring Asterisk to build feature-rich telephony systems for small and medium businesses.

1. Install, configure, deploy, secure, and maintain Asterisk

2. # Build a fully-featured telephony system and create a dial plan that suits your needs

3. Learn from example configurations for different requirements

AsteriskNOW

A practical guide for deploying and managing an Asterisk-based telephony system using the AsteriskNOW software appliance

Nir Simionovich   PACKT

AsteriskNOW

ISBN: 978-1-847192-88-2   Paperback: 204 pages

A practical guide for deploying and managing an Asterisk-based telephony system using the AsteriskNOW Beta 6 software appliance

1. Install an Asterisk-based telephony system fast

2. Build an office PBX using AsteriskNOW

3. Learn the AsteriskGUI web management interface

4. Configure IP phones and connections

5. Configure and use the conferencing system

6. Write your own applications for Asterisk

Please check **www.PacktPub.com** for information on our titles

PUBLISHING

Asterisk Gateway Interface 1.4 and
1.6 Programming

ISBN: 978-1-847194-46-6   Paperback: 200 pages

Design and develop Asterisk-based VoIP telephony
platforms and services using PHP and PHPAGI

1. Develop voice-enabled applications utilizing
   the collective power of Asterisk, PHP, and the
   PHPAGI class library

2. Learn basic elements of a FastAGI server
   utilizing PHP and PHPAGI

3. Develop new Voice 2.0 mash ups using the
   Asterisk Manager

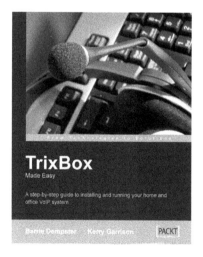

TrixBox Made Easy

ISBN: 978-1-904811-93-0   Paperback: 168 pages

A step-by-step guide to installing and running your
home and office VoIP system

1. Plan and configure your own VoIP and
   telephony systems

2. Setup voicemail, conferencing, and call
   recording

3. Clear and practical tutorial with case study
   format

Please check **www.PacktPub.com** for information on our titles

www.ingramcontent.com/pod-product-compliance
Lightning Source LLC
Chambersburg PA
CBHW060523060326
40690CB00017B/3369